WATER

Y0-ABP-498

EDITED BY PETER CARVER

WATER

PETER MARTIN ASSOCIATES LIMITED

Introduction

In *Water*, as in the three companion volumes in the Elements series, a rich blend of voices from every part of the country speaks of authentic Canadian experience. Again the contributions come both from writers with national reputations and from those whose names are not yet so well-known.

To Canadians, water has been a determining factor in all its forms—oceans, rivers and lakes, snow and ice, rainstorms and floods—and, indeed, in its absence, as in times of prairie drought. These are influences which have governed the contours of our communities, our industry, our recreation and our lives.

The word "water" evokes for most of us a string of personal associations going back to earliest childhood. The process of making the selection for this book has touched off my own chain of recollection. . . .

Snapshot: in the lagoon, sheltered and calm, turtles surface and blink in the afternoon sun, and small children lean over the gunwales to haul dripping long-tailed water lilies into the rowboat. Another: the northwest wind whips up chills of fear as an outboard, suddenly too small, is battered in the open choppy stretches of dark water. Again: a canoe glides out along the path of the August moon.

It has been seventy years since my grandfather —woodsman, writer and missionary—began weaving the family lore which still surrounds our island of rocks and trees near Kenora in northwest Ontario's Lake of the Woods. Indelibly associated with first memories for three generations, "the lake" is a metaphor for the flow of life.

I am a passenger in a train which dawdles beside the Columbia River, from Revelstoke to Arrowhead where the river widens to become the Upper Arrow Lake. Finally, after taking three and a half hours to cover thirty-eight miles, we puff alongside the dock at Arrowhead, a cluster of dark red railway shacks.

There the *S.S. Minto*, last of the lake's sternwheelers, waits for us. In some ways reminiscent of Stephen Leacock's *Mariposa Belle*, she moves out onto the smooth surface of the water, the great wooden paddles splashing rhythmically. On occasion she runs her shallow snub-nosed bow up onto the beaches of isolated lumber camps. Or she nudges up against the wharf of an incongruously large hotel, once part of a CPR resort project, now left in the hands of an elderly but wiry proprietor who is also porter, cook and guide.

The trip ends for me in the early evening at the village of Nakusp, with the sun slipping behind snow-capped Saddleback across the lake. Relatives are waiting at the dock to drive me in their Land Rover to the farm nearby where I will spend the summer browning in the hayfields.

More than twenty-five years later, a highway has replaced the train and the *Minto*. Now the shoreline of the Arrow Lakes is a grotesque confusion of mud-flats and half-submerged tree stumps, courtesy of the Columbia River power project.

An early morning departure on the Vancouver to Victoria ferry, wrapped in white Pacific fog. The water is barely visible from the ship's rail. Slowly, as we move away from the shore in a kind of dream, the fog lifts to reveal hazy vistas of islands and mountains, shafts of sunlight spotlighting patches of the flat sea, fishing boats puttering in the distance, gulls swooping close to snatch crumbs and galley refuse. The regular passengers see none of this, dozing or enfolding themselves in morning newspapers. But, as a first-time passenger, I must be everywhere at once, to drink in every detail.

A mile of white sand beach joins two rocky points on the south shore of Nova Scotia. Across a deep bay, the next low headland reaches out to the lighthouse, last rock until Portugal.

As I stroll out onto the beach from the woods and beach grasses behind, I am dizzied by the sudden change in scale. I walk along the beach at low tide, pausing for snails centimetring their way to some tiny destination, peering into rock-crotch tidal pools or deep clefts where anemones hide. Sandpipers patter their intricate water's-edge dance searching for microscopic snacks. Gulls drop their catch onto the hard wet sand, leaving a debris of purple-fleshed sea urchins and crab fragments strewn along the beach. Where freshwater meets saltwater is the richest of nurturing grounds. Seaward the waves cast up lobster traps and buoys, fish boxes, bits of mysterious skeletons, tangles of neon-coloured ropes and nets—treasures at every step. Several hundred yards out to sea where surf breaks over a cluster of black rock, three long-necked cormorants stand guard, silhouetted against the blue.

These images carry a strong personal significance for me. I hope the readers of *Water* will discover their own stream of associations as they read the selections that follow. Perhaps they too will refresh their sense of water as the element, more than any other, which surrounds us and nurtures us from the first moment of life.

Contents

The Creation of Man

After the flood, when the waters that covered the earth had receded, Raven walked upon the beach, disconsolate and alone, his mind a Wurlitzer of memories and old movies. Each familiar object and place set off a kaleidoscope of tricks and conversations, wild chases and indecencies among the rivers and islands of the northern coast. At first he laughed aloud as he recalled instances of his own wit and cunning; then, as the number of empty days mounted and the images from his scandalous past faded, Raven's face grew long. He drew his wings tight about him and hunched into the wind, the same wind that had soothed and buffeted him in the days before. . . . Before what? In truth, he hardly remembered.

So despondent had he grown that the simple task of feeding himself seemed odious. Back and forth he staggered along the narrow peninsula, shaking the fine grains of sand from his bloodshot eyes. He recollected, ever so faintly, a scene in which Humphrey Bogart dragged his scorched remains through the white sands and blinding sunlight, visions of water and home alternating before his eyes. Raven's eyes brimmed with emotion. He told himself it was the sand that irritated them. Then he withdrew into the shelter of an enormous cedar stump, beached dry above the hightide line. He drove his beak into the black feathers under his wing, but he could find not even a single flea or louse.

As the tide ebbed, Raven's hunger overcame his feeling of despair. Leaving his shelter, he plodded to the water's edge and began to examine the moist sand. He cocked his head in the manner of his former friend the robin and listened. Pleep . . . pleep . . . pleep. Raven edged closer to the sound. It came again. Poleep . . . pleep . . . pleep. Raven put his ear to the ground and strained to hear. Suddenly a spout of water erupted from the sand, hitting him square in the eye. Raven stood aside, shook off the salt water, and made a mental note that he would call this spot Sand Spit Point in recognition of the practical joke that had been played upon him. Then he began to dig furiously with his talons in the sand.

Within minutes there lay at his feet two beautiful blue-grey clamshells, one the size of a modest hamburger and the other no bigger than an egg. What was unusual about them, aside from the colour and the strange noises they emitted, was the fact that they began to grow before Raven's very eyes. Soon the larger of the shells was the size of a cowpie and the smaller had reached the hamburger stage. Raven was aghast. When the larger clamshell became the size of Raven himself and the lesser clamshell had reached his knees, Raven witnessed an astonishing sight. Of their own accord, the lids of both shells opened slowly, all the time emitting the same strange sounds that had first reached his ears. In one shell lay a child, curled, knees to its chin, in fetal position, eyes not yet open. Its colour was not white like the sand, nor dark like Raven, but brownish-red like the arbutus. Raven looked on in wonderment.

In the other shell, more wonderful still, was a woman, full-blown and voluptuous, eyes also closed and the smile of the blessed upon her lips. Raven stepped back and let out a groan that startled the wind and which to this day is still to be heard by the chosen ones on Sand Spit. His chest swelled, his eyes rolled like beads in a rattle, his feathers fluffed out in blue-black splendour. Quickly, as if to snatch the dream from its vanishing, Raven made up his mind. Gathering the child in his wings, he took the waking woman by the hand and led her down the beach and into the forest, which closed gently around them, leaving only a faint chuckle on the wind.

Thus did Raven's marriage with the sea create the Indian peoples of this coast.

Haida legend adapted by Gary Geddes

All the Diamonds in This World

All the diamonds in this world
that mean anything to me
are conjured up by wind and sunlight
sparkling on the sea

I ran aground in a harbour town
lost the taste for being free
thank God He sent some gull-chased ship
to carry me to sea

Two thousand years and half a world away,
dying trees still will grow greener when you pray

Silver scales flash bright, and fade,
in reeds along the shore
like a pearl in a sea of liquid jade
His ship comes shining
like a crystal swan in a sky of suns
His ship comes shining

Bruce Cockburn

Under the Top of the World

During 1974, Dr. Joseph B. MacInnis led a team of five Canadian scientists and photographers on man's first underwater exploration of the North Pole. This is his exclusive account of that adventure.

The ice overhead shivers with brilliance. An infinite skylight bathed in sapphire, with menace hidden behind its beauty. I hang suspended in the crystal silence; below me the ocean plunges 13,000 feet into black oblivion.

I burrow down inside my wrinkled diving suit. A trickle of flame-cold water scalds my neck. Ahead, in the gloom, is an enormous white wall—a tortured construction of gaunt, bare slabs of ice. At last, after four years of planning, I am under the North Pole . . . the top of the world . . . a white ghost place where all meridians meet and every direction is south. A wild unmarked spot where it is midnight for eighty days and the midsummer sun never sets.

The dream of being the first man to dive under the North Pole had been with me since 1970, when I brought my first expedition to the Arctic and spent seven days exploring under the ice of Resolute Bay. We returned to the Arctic twice for more dives, always building up to this supreme adventure under the Pole, one that we knew would challenge our courage and our physical endurance to their limits. Then, at the last moment, it appeared that all our efforts and all our preparations would end in failure.

For days, bad weather made it impossible for our pilots to take off on the final 450-mile flight to the Pole. When the weather finally cleared all we could see below us was hundreds of miles of tortured ice at least two fathoms thick, impossible for us to cut through in the three days we had remaining for our expedition. But as our hopes dwindled, our pilots located a narrow lane of thinner ice right at the Pole and set us down to build our campsite.

Now we have punched through and I look up as the black figure of Rick Mason, an underwater photographer and my companion on three polar expeditions, slips through the dive hole. As he moves toward me with rippled grace his exhaust bubbles rise like a small flight of transparent birds and come to rest on the pale undersurface of the ice.

He carries a large spotlight which drives yellow brilliance through the water. Its thick black power cord snakes out behind him and up to the dive hole as he begins a downward arc toward the ice wall.

The wall was formed as the frozen waters rammed together in an agony of buckled and broken slabs. Propelled by huge and distant winds, the ice has been forced downward into a pressure ridge of fractured terraces. All is deathly silent; nothing hints at the tumult of formation. The wall glows with a heatless light that appears incandescent, a light naked of life and stripped of warmth and safety.

I follow the slow churn of Rick's black fins. As the white cliff draws closer, batwings of apprehension flutter in my chest. I chase them away with thoughts of the magnitude of our being here and the purpose of our adventure.

Knowledge surely. A better understanding of the forbidding ice world that is an important part of Canada. We are trying to explore and comprehend an unseen portion of the planet.

We are also here to better understand ourselves. Men looking for personal limits—each one of us asking the lonely question of self-appraisal—can I? Al Purdy put it well when he wrote:
"I'm so glad to be here
With the chance that comes but once
To any man in his lifetime
To travel deep in himself
To meet himself as a stranger
At the northern end of the world."
We are here to test ourselves and our equipment. How long can we stay? What life-support do we need? What techniques must we develop? This is the most hostile environment on earth and answers learned at the North Pole will widen the gates into other parts of underwater Canada.

Something feels wrong. I look down past Rick's floating form and see the flutter of falling silver. I reach for my camera, but bulky rubber fingers only grope against contours. My viewfinder! A $300 wide-angle lens is disappearing into midnight. I race down, but my plunge is useless. The lens falls too far, too fast. It is well into a two-mile journey to the bottom. I gasp for breath and wonder how to explain the loss to the friend who loaned it to me.

Rick now plays the light beam over a series of sharp angular blocks jutting out from the main bulk of the pressure ridge. Shrill whiteness turns splendid yellow. Several of the uppermost slabs are covered with sparkling ice crystals.

The light enhances our courage and we slip slowly down toward the winter twilight under the

ridge. Columns of crushed ice run in chaotic lines in all directions. At an unknown point in the gauzy light the enormous white keel descends into the frozen indifference of the sea.

We are almost fifty feet down. The dive hole has faded to a small white square. The keel extends a fat bleached lip of ice out toward our feet. We drop lower for a look beneath it. As we descend our shoulders nearly touch. Our paired breathing lifts away long sighs of near-freezing air. Together our eyes probe the darkening waters. The skeleton shape of two deep ghost peaks stare back. They are a long swim down, further than our courage will take us.

Somewhere below the blackness is the sea floor. On it are mountains standing 7,000 feet high. They are part of a range that begins in Labrador and runs for 2,000 miles up the eastern shore of Baffin and Ellesmere islands. At the northern tip of Canada these mountains plunge into the sea to become the Lomonosov Ridge. They end their long undersea journey in Russia as the New Siberian Islands.

We drift alone in an enormous body of water. The Arctic Ocean covers some five million square miles; it is bounded by the frozen coasts of Asia, Europe, Greenland and North America. We are 450 miles from the Canadian shore.

The light swings in a slow wide arc. Its glare runs up and over a staircase of torn and tilted blocks. I picture the stormblast of creation: the pack ice in slow movement; howling dark winds far to the south; a groaning collision of ice massing high upon itself; the screaming pressure of blocks grinding deep into the sea; the final mute stability of countless interlocking fragments.

The wall appears anything but stable. It slopes up and away from us. A giant purple-blue overhang. The buoyancy of ice exerts tremendous upward force. Several of the larger blocks are massively undercut and look ready to break for the surface. I picture tons of ice roaring skyward in a runaway upward avalanche that would mean instant death.

We delude ourselves that death will not come if we carefully evaluate the hazards of the sea and the techniques to be used. But diving—especially under the polar ice—is never absolutely safe. Our awareness of this fact is an iron fist, heavy on the shoulder. My arm and hands ache from the cold. Rick and I nod and pull back from the darkness.

Through the mirrored surface of the dive hole I see the distant specks of our three companions on the surface. Suddenly, just ahead of my face mask, a sliver of glass winks and is gone. A faint animal flicker, a tiny scrap of life against a chaos of tortured ice.

Rick and I pause just beneath the surface. From here we can see the full writhing axis of the pressure ridge. It is an endless chain of suspended and unconquered peaks. Canada's oceans contain thousands of miles of similar ridges, but few men have had this intimate underwater view.

Little is known about the architecture and geography of pressure ridges. Knowledge has always been obtained indirectly or secondhand, but now the technology exists to allow scientists to make firsthand observations.

The two of us move into the white water immediately under the dive hole. I see the harsh chisel marks made as the surface team cut through the ice. The three of them, Ernie McNabb, Gary Holmes and Patrick McLaren, lean over to watch us and help us from the water. McLaren is fully dressed and ready to dive. He's a geologist and anxious to get down to the pressure ridge.

On the far side of the dive hole a small Canadian flag waves in the current of our arrival. The flag is perhaps the best expression of why we have come. The North Pole is the northern apex of Canada's ocean territories. If we, as a nation, are to understand and manage these territories we must be able to operate anywhere on them—including above and below ice-covered waters.

A hand reaches out and my head breaks through into the cold polar air. Rick emerges, lifts his face mask, and through a beard instantly coated with frost, exclaims: "That was the damnedest experience ever!"

Joseph B. MacInnis, *Maclean's*

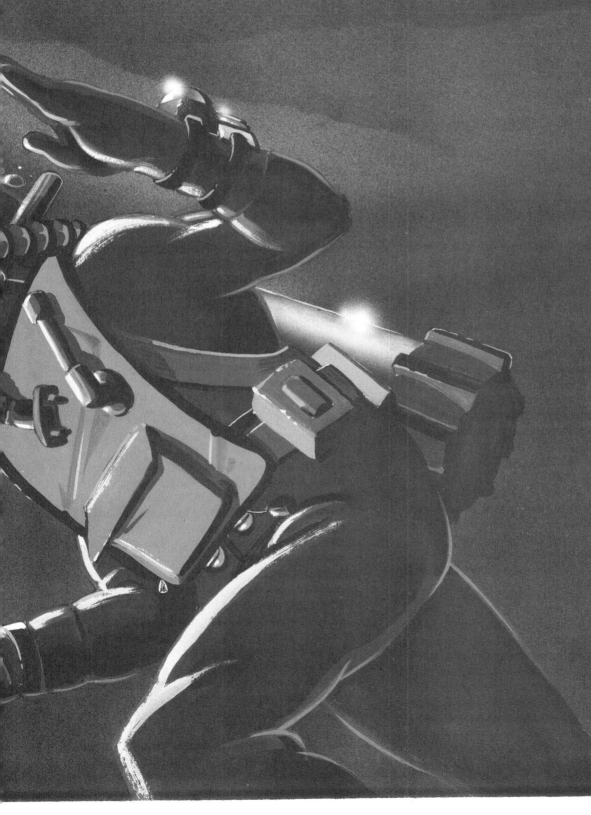

Further Arrivals

After we had crossed the long illness
that was the ocean, we sailed up-river

On the first island
the immigrants threw off their clothes
and danced like sandflies

We left behind one by one
the cities rotting with cholera,
one by one our civilized
distinctions

and entered a large darkness.

It was our own
ignorance we entered.

I have not come out yet

My brain gropes nervous
tentacles in the night, sends out
fears hairy as bears,
demands lamps; or waiting

for my shadowy husband, hears
malice in the trees' whispers.

I need wolf's eyes to see
the truth.

I refuse to look in a mirror.

Whether the wilderness is
real or not
depends on who lives there.

Margaret Atwood

Image

We watched the canoe go from us. The sun, low over the black rim of the western mountains, slanted on the lake waters until they became a carpet of creeping flame, failing as it advanced towards us, until at our feet only black water lapped, cold and spent and sobbing in the sandy runnel where the canoe's prow had rested. The canoe was wide, sat low in the water, two pairs of figures—four of Dobble's men—stooped monkishly against its centre thwarts, their backs to the west. The shape of the paddler in the stern rose above them, paddle flashing sword-like from the water and streams of water, blood-reddened against the sinking sun, running from its blade. For a long time we heard the tinkle of those falling paddle streams and the widening wake as a sigh upon the flaming waters. Then the four heads and the paddler's back and the canoe merged, blurred, became black and small and still, consumed before our eyes in the fiery expanse of lake and sky. As our vision faltered, the paddle flashed again, the lake's red bosom rose and swelled, and on it the black speck diminished to a quivering point of dissolution, hesitating one final moment at the fire-guarded gate of the world's end. Then we heard the canoe touch the other gravelled shore. We heard men's voices, and before us the sunset rode triumphant on the waters.

from the novel *Tay John*, by Howard O'Hagan

On the Wanapitei

The Wanapitei River flows out from the lake of the same name and crosses the Canadian Pacific Railway line about sixty miles from the lake. Our purpose was to make the lake and from it reach the headwaters of another water system farther west.

Fine and fit I was and, as I thought, ready to deal with any kind of water.

We put our canoe into a large pool at the bottom of a rapid and proceeded to edge our way up to the foot of the rapid, whence we should make our portage, saving a carry of several hundred yards, always an important consideration.

"Now take this carefully," instructed my chief. "See that point jutting out? Well, immediately round that point we strike swift water. Take it quietly, then hold her steady till we catch the eddy which will carry us right up to the foot of the rapid."

It seemed quite simple, and it was to those who knew its secret.

We paddled quietly to the point, then pushed the nose of our canoe into the swift water; immediately there seemed a call for a fight. Desperately I dug into the swift water, paddling for dear life.

"Easy!" sang out the chief, but I had missed the eddy and we drifted downstream, then gradually edged across the river once more to the bank far below our starting point.

"Now you'll learn how to take swift water," said my chief with a chuckle. Again he carefully explained the technique of making an eddy help you against swift water:

"Put her nose into the current quietly, hold her with an easy stroke till I swing her stern about, then when the current strikes her put your back into it and she'll walk right upstream. Don't work hard till the stream begins to work with you, wait for my word and then pull like blazes."

It was all a mystery to me, but I determined this time to wait for orders.

Once more we approached the point still in quiet water, pushed her nose out into the current.

"Steady! Hold her!" Gradually the stern swung about. "Now!" yelled the chief. A few quick hard strokes. She held her nose upstream till the current caught her sidewise, then gradually she began to crawl slowly upstream and in two or three minutes we were beyond the point, caught by the upstream swing of the eddy and were soon being carried easily up to the very foot of the fall where we made our landing safely for the portage.

"Nice work!" said my chief. "Got it just right."

I stood looking back on our course in utter amazement. It seemed a miracle.

"How did that happen?" I asked.

"Quite simple. Immediately you pass the point, there is a backflow of the current close inshore. Catch that and the eddy pulls you right upstream. At the foot of almost every rapid there is an eddy on one side or other of the river. Get her nose into that and up she walks."

"She's a wonder!" I remarked, gazing at our canoe with new awe and respect.

"She's like nothing else that floats. She's a wonder but you've got to know her ways. You can't bully her."

I gathered that it was only one more mystery associated with things of the feminine gender. The main thing to remember, apparently, is that they will not stand bullying. A most useful addition to my psychological education.

Three days of strenuous toil fifty miles upstream—and a fierce stream, too, with twenty or thirty lift-outs and portages as it seemed to me—brought us to a large and perfectly enchanting lake, Lake Wanapitei. Next morning at dawn to avoid rough water we crossed the lake, a half day's paddle, and found a little stream flowing into the lake from the opposite side.

"This looks like the trail," said my chief, studying the map which he had from a Hudson's Bay Company agent. Up the little stream we pushed our way through tangles of bulrushes, willows and log jams and found ourselves at the edge of a huge swamp, hemmed in on three sides by a rampart of rocks.

"This is what I wanted to see," explained my chief. "We will make camp, snatch a bite of lunch and explore these ledges."

We spent the afternoon working our way round the rocky ledges, collecting what seemed promising samples. As the sun was beginning to make its descent toward the treetops of the swamp below us, my brother suggested:

"You work your way back along this ledge. I shall go around to the other side, then cross the swamp and meet you at camp." His plan was to follow the arc of rock around to the farther horn of the arc and then, following the string of the bow through the swamp, meet me at our starting point.

Slowly I worked my way back till I approached our meeting point where I sat down on a luxuriant bed of brush to wait his coming. After a wait of an hour or so I began to listen for any sound of his hammer, but could catch none. I could see no sign of him on the ledge about half a mile away across the swamp below me. I raised my voice in a loud call, but there was no reply. Once more I sent forth a long, loud coo-e-e. Far away across the swamp I caught his answering call. A little later I called and again got his reply. He was evidently crossing the swamp toward me. I moved toward our meeting point, called and got his reply just below me in the swamp.

"I'll go down and get the fire going," I called and proceeded to our camp by the stream. I put on the kettle and climbed the hill to meet him. But there was no sign of him. Again and again I called, but no response.

"Hurry up, Rob, the kettle is boiling," I said. But he paid no heed. What could he be doing? There were no rocks down in that swamp. Was he trying to fool me? That was not the kind of thing he would do on a trip like this.

"Hurry up, Rob! Don't be an ass," I said impatiently. "I'm going down to fry the bacon. Hurry up!"

Annoyed at his silly silence I went back to the fire, cut the bacon, got the fish ready for frying. Still no sign or sound of him. Once more I climbed up the bank and called down into the swamp, "What's keeping you? Hurry up! Everything is ready!"

Immediately at my back, about a dozen yards away it seemed, among the underbrush, there was a wild cry like a scream. Then I was properly angry.

"Look here!" I said. "If you think that's smart I don't. Don't make a fool of yourself. I'm going to have my supper."

So saying I went down to my fire and began to fry the fish and bacon, expecting him every moment to appear. I called, still no answer. Had he forgotten a hammer or something and gone back for it? Could anything have happened? I became anxious. Then I raised my voice in a long, loud halloo, when far across the swamp I heard his voice in reply. Again I called and again came his answer. At length he emerged.

"Well," he said pleasantly, "supper ready?"

I was perfectly furious.

"What sort of confounded idiot are you anyway?" I said. "Do you think that sort of thing is funny? How often have you told me never to fool in the woods?"

"What's the matter with you?" He gazed at me astonished.

"How often have you told me that no one but a darned fool would play that kind of trick? What did you go back for? Anyway, you might have explained to me."

"I don't know what you're talking about, Charley," he said, scanning my face curiously.

"Do you mean to say you weren't here fifteen minutes ago?"

Then I told my tale.

"By Jove, boy, it was a cat, a lynx. And how near did he come?"

"He yelled at me from that clump of bushes right there!"

"Good Lord! You are lucky, boy. He might have dropped on you from a limb overhead. It was your fire likely scared him off. They are inquisitive devils."

"But it was your very voice, I would swear to that. Not once or twice, but a dozen times. He answered me just below there at our feet. And then after I came back from putting on the bacon, he screamed at me from that clump. It was exactly your voice." Then I was enraged at the beast. I got my revolver and beat round about and through the bushes but could find no sign of my clever ventriloquist.

That night beside the fire Robertson told me of similar experiences he had heard of from one of his trapper friends.

"But I always thought they were pulling my leg a bit. The brute can imitate any cry or call, they say."

"Well, he certainly fooled me," I declared, "even to that last wild scream. That, of course, was like no human voice. Still I thought it was you trying to startle me. But all the way across that swamp as he came nearer and nearer I never for a single moment doubted it was your voice answering me."

I slept with my gun under my pillow, hoping my friend might be inclined to pursue his investigations. But evidently he had satisfied his curiosity.

We penetrated the wilds for another week; failing to make connection with the waters of the western watershed, we were forced to retrace our way over Lake Wanapitei. We determined to make the run down the river, about fifty miles, with its two dozen lift-outs and portages, in a single day. It was all swift water, tricky and not without danger, but the chief had been down and he had a mind that registered a course like a map.

At dawn we were away. Pausing twice only to boil our kettle, our grub being all finished, we found ourselves in the evening at the head of the last long sault. The question now was a run or a carry. The carry was long and up over a rocky climb. We should have to make two trips. It was not a pleasant prospect. We were faint from lack of food and tired out with our hard day.

"Let's take a look at her," suggested Robertson.

We climbed up to the top of the rocky bank and looked down upon the rushing water, with its long, dark, smooth sweeps and its sharp pitches of white water. Long and carefully he studied the course.

"She looks rather wicked," he mused. "That centre chute is a bad one. But if we hugged the left bank we could avoid that chute. That is really the only bad pitch—and she's a devil. Of course I believe we could hug this left bank all right."

"Suppose we couldn't—suppose we were drawn into that centre chute, could we make it?" I asked.

"We could, but it would be a close thing. You see that long smooth sweep there just before the drop. If we hit that it would be Kingdom Come for us. But if we did strike that centre chute fair a good heavy lift to the right would clear the boulder all right. It can be done. What do you say?"

"Whatever you say," I replied, hoping he would try it.

His eyes were beginning to shine. He glanced at me with a curious searching look. I grinned back at him.

"By Jove, we'll do it!" he said with an answering grin. Again he scanned the course foot by foot, studying the water and explaining it to me.

"If we keep close to the left bank the thing is straight going. But there is a heavy cross current there toward that centre chute. Now listen. If we do get caught and are drawn in, a quick heavy lift to the right just before we reach that chute will send us clear. Be ready. If I yell, then lift like blazes and don't lose time. Sometimes a split second of time means eternity."

As I walked back to our landing I was conscious of that curious thrill through my nervous system that I used to get just before the kickoff in a big match with McGill, but with a deeper and more portentous interrogation in it. We made our load snug and taut.

"All right, boy, get in!" His voice had a kind of gay lilt. I got in, kneeling low and spreading out my knees to grip the sides of the canoe. A touch of

the stern paddle and we slipped smoothly into the current. At once I became conscious of a quick, strong grip on the canoe, as if someone had said, "Come on here! I've got you! You are mine now!" For twenty-five yards we kept gathering up speed, not by seconds, but by fractions of seconds. We are at the crest. Jove! This looks like a toboggan slide! And feels like it too! We are flying as if in air, a curl of white water smites us in the face! She begins to leap, to plunge, she skips from one wave crest to another, pounding only the tops of these hummocks of water that feel like rocks under us. Suddenly I see that we have missed the current. We are in the centre—the big plunge is rushing at us. "Li-i-ft her!" The yell smites like a blow on my senses—what I did I don't know—a wash of water chokes us—blinds us—our craft pitches madly forward, exactly like a bucking bronco—we are under—we are out into the air—gone—no! I am sitting in water over my knees—but—yes—we are through and skimming over crumpled, broken, boiling wave crests—but we are through!

"By Jove, boy, you did it!" gasps Robertson, exultant. "You're a whale."

A loud cheer from the bank greets us. It comes from a little group of men standing at the landing.

"Heh! What the hell? Oh, it's you! Might 'a' known it was some damn fool water hen!"

"Hello, Mike! Glad to see you, old boy," shouts my brother, bringing his canoe up to the little wharf.

"Say, you darned old fish hawk! You gave me the wolly-woogles, all right!" We shook hands all round, then Mike said solemnly:

"Look-a-here now, boss, you'd ought ta knowed better'n that. When I saw you up there at the top of that pitch I says, 'Well, there's another damn fool goin' to hell!' That's what I says, ask Bill here!"

"Yes, that's what he says, all right," replies Bill; slowly squirting a stream of tobacco juice into the river. "Two fellers, lumberjacks, drowned in that water this spring, comin' down in a Hudson Bay pointer too."

"Well, Bill, you see," said Robertson with a little grin on his face, "this isn't any Hudson's Bay pointer, and, by golly! we are not any lumberjacks. See that boy?" he added, shaking a finger at me. "He's—he's—well, look at him! Not much to look at—but you see before you the best bowman in this north country! Bar none!" Which I knew to be wildly absurd, but no less overpowering for that. Furthermore, it expressed his feeling at the time.

My throat was hot and choked. I couldn't utter even a croak. I made a jab at his stomach with my paddle and hastily bent to unload our stuff. Gold medals? You could have them all. But that word, from perhaps the finest canoeman in that land of canoemen, knocked me out. I couldn't see what I was doing, but it took me some minutes to unload my stuff. Good old Rob! Miner, woodsman, canoeman, rifleman, none better in all the Northland. A leader of men who would follow him anywhere, no matter against what odds, and proud to do it. This was not my last trip with him, but never with him or with any other did Death splash water in my eyes as during those split seconds on the Wanapitei.

Charles W. Gordon

Notes on a Northern Lake

Rainy River Country, Northwestern Ontario

1.
We freeze in our beds
The wind sends shivers across the lake

2.
We swelter in the open air
The sun sends lava across the lake

3.
We move among the trees
And through the woods the lake moves too

John Robert Colombo

#37 - Free XA - Centre

Viewing Indian Pictographs at Bon Echo in Eastern Ontario

for Merrill Denison

We cut across the lake to the immense table rock which rises so steeply out of the water. Although the overall effect of the rocky surface before our eyes is grey, it reminds us of an abstract canvas. Lichens and minerals, rather than paints and dyes, have splattered and stained the greyish rock with vivid orange, purple, white and black.

"The best-preserved ones are over here," Merrill says, raising his voice over the noise of the engine. He steers the barge a little way, then cuts the motor. For the first time, as we drift toward the rock, we can hear ourselves think.

"There's a hand!" "A man!" "Look at the monster!" "See the little canoe!" Like children we point out the recognizable features to one another. On the face of the rock, in faded ox-blood, the pale markings, delicate but decisive, are many things.

"This is the best way to see them," Merrill says. He takes an oar and splashes the figures with water. He does it again and again, a clumsy sacrilege. But, well-weathered, the images are used to it. They suddenly glisten, come alive.

The Moose-Man battles the immense Bug-Monster, two legs pitted against four in some epic engagement centuries or eons ago. Twenty stick men, with staves over their shoulders, await the outcome from the safety of their canoes. The battle is never won. . . .

"There are other pictographs north of Lake Superior," Merrill says, "but these are the only ones so far south. They're at least two hundred and fifty years old, probably much older, nobody really knows. Somebody said they were 'dream images', and that the medicine man told the artist of the tribe what things to draw and where to draw them."

The sun begins to set and the air grows chill. The engine coughs and starts up again, but my thoughts remain with the Indians, the gentle people and their battle. They have come and gone and hardly hurt the land, and the land has hardly noticed or hurt them. Here are the few precious signs that another people have lived here before us. Their images have a meaning that we sometimes share.

"They look exactly the way they did when I first saw them. That was more than sixty years ago," I heard Merrill say over the throb of the motor, as we cut our way back across the lake. "They haven't faded at all, and nobody knows a thing more about them today than we did then."

John Robert Colombo

The Fire Canoe That Was Never Launched

Shivering in the anonymous hold of a New York-bound immigrant carrier, Tomi Jaanus Alankola dreamed of a promised land, a refuge from the economic chaos of nineteenth-century Finland. Born on September 23, 1878, in Koronkyla (a tiny settlement in the province of Vassa in western Finland) Alankola had learned the only trade available to him, and now, aged twenty and on the shores of America, he hoped to earn his living through his finely tuned skills in the art of shipbuilding. However, he was arriving amid the hungry throngs of workers from innumerable lands, with whom he would have to struggle to earn a living; and so he pressed inland, into Ohio, Michigan, Minnesota, and Wisconsin, searching for work merely to survive. He tried homesteading and mining, among other pursuits. He married and fathered a son and three daughters.

Here the story takes its first twist.

In the autumn of 1911, after perhaps a decade of labouring, Tom Sukanen (the name he registered in America) suddenly picked up and left. With no money to speak of, a growing family to care for, no business instinct to send him off on a wild enterprise, and no visible reason for his impulse, Sukanen set out, on foot, in search of a brother who lived in Canada. He carried on his back the basic necessities for the trip, and nothing else. How he found his way through the wilderness of badlands, vacant plains, and parched prairie brush is undocumented and unexplained, but a Finnish sailor who could navigate the open sea was not to be daunted merely by unfamiliar terrain. What made a sea of plains any different from the Baltic?

Whether by instinct or by luck, six hundred miles after Sukanen left his Minnesota homestead he crossed the final rolling hills onto the Macrorie-Dunblane pastureland of central Saskatchewan. Near the tiny hamlet of Birsay (some seventy-five miles southwest of Saskatoon) he found people who knew his brother, Svante. Sukanen decided to stay; on October 23, 1911, he filed entry for a homestead comprising the northeast quarter of section fourteen, township twenty-six, range nine, west of the third meridian, about a half-dozen miles west of Macrorie. To his neighbours, farmers like Vic Markkula, Herbert Fredeen, and W. A. Cahoon, Sukanen seemed just another immigrant farmer-settler who worked hard and made the most of each season.

His nephew, only a small boy at the time, heard Sukanen referred to as a "well-to-do" farmer. "I think he often worked out for his neighbours," suggested Elmer Sukanen, rather vague about all the facts surrounding his uncle's strange, almost completely forgotten story. "He used to go and help them stook and work on thrashing outfits. But he never visited us too much, partly because we lived ten miles apart; he was more of a loner."

Sukanen was also inventive, even creative, in the variety of gadgets he dreamed up and constructed, most of which benefitted no small number of the community's farming families. As Moon Mullin observed, "Tom made the first grain-thrashing machine, the first in the district . . . a power thrashing machine. He also made a sewing machine, so that women [of the area] could mend clothes . . . things that helped people. And then he showed a lot of people how to build homes for themselves when they first came. He was just a little different, more of a stand-out man than the average." Sukanen wasted nothing; during the mid-Depression, he knit himself a suit of clothes out of binder twine that "wore real well".

"Another time he had built a camera," continued Moon. "He went to visit a lady . . . to take her picture. The camera, I guess, burned magnesium. Tom arrived and he asked her to step in to the doorway, and of course she didn't know what was going on. . . . There was this terrible explosion . . . and Tom, he disappeared over the hill, his outfit in hand. But two weeks later he came back with this beautiful photograph to show her how that machine that he'd made worked. That was the way he was. He thought he could do these things, and he went ahead and did them."

"I can remember that bicycle, too," added Elmer Sukanen. "'cause I used to fool around with it all the time. You didn't have to pedal it; you pulled it like a hand car. I was about eight or nine years old," continued Elmer. "I can remember his place, which resembled a silo built on a hillside with a staircase running through it and a mirrored periscope mounted on the top for looking around the countryside. And he had this car rigged up. He wouldn't start it because once he had the doggone thing in gear and he had a crank in front. He flipped it and it took off; damn near run him over. So next thing he went and rigged up a crank beside him so he could crank it from inside the car. That's how I remember him. And he had a violin he'd made and he played it, although I never heard him play."

By May of 1916 Tom Sukanen had received patent for his homestead, and by prairie (and

wartime) standards, had succeeded in homesteading and farming—as a grain-grower, livestock raiser, and farm manager he had amassed approximately nine thousand dollars, an enormous sum for those days. And still a wife and four children waited patiently a half thousand miles away on another homestead for Sukanen to return. In 1918, after nearly seven years with no communication to or from his family, Sukanen walked the same six hundred miles to Minnesota; he was full of a new spirit of triumph and news of the home he'd build for his family. On his return he found no one on the homestead. His wife had died of influenza and his three daughters and one son had been scattered to the wind by local authorities into various foster homes. The boy was the only traceable member of his family; his name had been changed to John Forsythe. Sukanen gathered up his son and started home, only to be caught south of the Canada-United States border by the authorities. The boy was returned to the foster parents, but lived in expectation of another planned escape with his father. Authorities prevented John's second escape attempt and immediately threw the boy into a delinquent orphan home, and expelled his father from the United States.

The Crash of 1929 spread ruin throughout the prairies. Everywhere was chaos and collapse. With depression surrounding him, both economically and mentally, Sukanen began to draw in his mind's eye a bizarre, almost incredible plan, unlike any other in the history of prairie steamboating, that was uniquely his own. In the dust of those Dirty Thirties, he would build a steamboat on his homestead which he would drag seventeen miles to the banks of the South Saskatchewan River, and then sail down the Saskatchewan to either the Nelson or the Churchill River, through Hudson Bay, past Greenland and Iceland, and home to his native Finland. Of shipbuilding, navigation, and marine lore, Sukanen still remembered much from his youth spent among Finnish fiords; and for courage and determination he was unsurpassed. With the decision made, Sukanen considered the expedition as good as complete. *Dontianen* was born in that vision.

The first evidence of unusual activities in the northeast quarter of section fourteen came in the shape of large quantities of sheet steel, cable, copper, and other such supplies brought in from Port Arthur at the Lakehead. Sukanen began sinking his entire life's savings and every ounce of energy he could spare from his farming into the construction of the steamboat. Soon his sowing and

reaping would be left aside in his obsession to complete and sail the vessel. As soon as materials arrived at the Macrorie railway station from the eastern factories, Sukanen would collect them and, with the help of a neighbour, haul them to his farm.

Dontianen (a Finnish word meaning "small water bug") rose in three sections: a keel, a hull, and a superstructure, the latter consisting of a cabin, a wheelhouse, railings, and other trimmings. The design resembled a mid-nineteenth-century Scandinavian cargo freighter that could be converted conveniently from sail to steam power at will. First to take shape was the hull, some forty-three feet in length and thirteen feet at its greatest width, and ten feet from keel to deck. Once Sukanen had positioned the ribs of the hull which formed the main frame, he then shaped the exterior of her body with lapped planking, which he both tarred and caulked. Beyond this sealed hull covering he secured a second layer of non-lapped planking, and then applied the final protective covering of one-sixteenth-inch sheet steel, which he shaped and cut by hand with his self-designed tools at a forge he built expressly for the purpose. On each sheet of the outer steel shielding he crimped the edges in order to interlock the pieces of steel for greater strength. He intentionally left the deck wide so that the steam boilers and engine mechanisms could be lowered into the heart of *Dontianen* at a later date.

The North American steamboat's grandfathers (the Mississippi keel boat and the Northwest Territories York boat) usually had flat-bottomed shallow-draft hulls to facilitate passage across the sandbar and snag-ridden shallows of the mid-western waterways. But Sukanen knew that an ocean-going steamship would require a keel to keep her upright and stable in rough currents and gale-force winds. Ballast and freight could be taken on in the keel as well. Nine feet high and nearly thirty feet in length, the keel was double-planked, as the hull had been, then tarred, and then enclosed by sheets of galvanized iron, which Sukanen laced together with unbroken steel wire. This way the keel would be durable enough to support *Dontianen*'s superstructure, and would also be flexible on the high seas. Sukanen then sealed the keel section by smearing it with horse blood, a process he calculated would protect the underside from the corroding effect of salt water. By preparing each section of the steamboat separately, he hoped to insert the hull section into the keel, double-boiler style.

To crown these lower stabilizing sections of

Dontianen, Sukanen designed and built two eight-foot-high cabin arrangements, each with four-foot-tall railings to be trimmed with some of the softer metals he had purchased from the East. While the one cabin toward the bow was the wheelhouse (enclosing the steering mechanism), the second cabin, astern, would provide living quarters and the space to store needed supplies. The upper deck quarters would not only hold navigational instruments and a unique water-clock chronometer, but also cupboards, bunks, and living space. Heat from the wood-fired steam engine and the smokestacks would pass through or near the cabins on deck, as the heat was expelled. Sukanen also forged—by hand, from solid pieces of flat steel—pulleys, gears, funnels, a propeller with driveshaft and universal joint, a lifeboat, and numerous chains. Incredibly, this was all done while Sukanen was in his late fifties, and during the Depression, when most prairie people were concerned with bare survival.

What was Tom Sukanen like? Though few of his neighbours were in accord about his psychological stability, most agreed regarding his physical appearance: chiseled facial features; skin stretched taut over a symmetrical face; blown, unkempt coal-grey hair; wide, expansive forehead; square chin; deep-set eyes; indistinguishable neck; dominant, muscular shoulders; tall, massive frame. "He was a giant of a man . . . because men six-feet-two, when they talked to him, they looked him right in the eye. . . . And his chest was a lot larger than an ordinary human being . . . his strength, the strength of three men . . . and he often proved it."

"He was a husky guy," recalls Elmer Sukanen. "His hands were all bloody always, 'cause he was workin' with barbed wire barehanded."

The small-community nature of the West, combined with the conditions of the Depression, exaggerated Sukanen's stature and his unconventional behaviour out of all proportion. Fear is not too distant from amazement. Macrorie district town and farm people saw the man as "different", all but condemning Sukanen when they labelled him "superhuman". "No man on earth could put up them big planks and steel like he did, unless he was a superman in strength. And his chest expansion, well . . . a man told me he himself had been a big man, and yet when he put on Tom Sukanen's coat one day, he said, 'Lord,' he said, 'it hung on me like an oversized overcoat.' That same man weighed 270 pounds when he was younger. I know. I seen him and I can imagine. And at 270 pounds Tom's coat was still big on him."

Another man, named Stone, saw Sukanen perform a remarkable feat of strength. Two railway workers were lifting a set of train wheels. As Stone remembers, one of them said to Sukanen, "well, you're a big strong man, you get on the one end." And instead of getting on the end, Sukanen walked over and picked up the wheels in the middle and put them above his head. Then he just threw them out and let them fall. The railway workers figured afterwards the wheels weighed six hundred pounds. The two workers, according to Stone, were among the biggest men in the district, but even using tools the two of them couldn't lift the wheels.

For six years, day and night, summer and winter, without a single interruption for relaxation or illness, Sukanen drove himself fanatically to complete his ship. The keel and the hull neared completion at his farm, the superstructure took shape at the river's edge some distance away, and all along the route between the homestead and the river, portions of *Dontianen* were spread about. Sukanen lived in the overturned hull or in the cabins of the superstructure, depending on where he was working at the time. Still arriving from eastern factories was the sheeted steel Sukanen rolled into pistons, shafts, rivets, a smokestack, cylinders, boilers, valves, pipes, and a wide assortment of cabling and rigging. Throughout at least one winter he worked night and day over his forge, cutting the boiler, the pumps, the propeller, the gears, and the other steam-driven mechanisms out of shapeless steel.

The arrival of spring some six years after *Dontianen*'s birth redoubled Sukanen's obsessive urgency. He was approaching sixty years of age, and inevitably his strength was flagging. His neighbours doubted his sanity. His spirit was beginning to break, in part through the recently arrived news that John, his son, the only surviving member of his family, had died. And not only did *Dontianen* have to be transported those seventeen miles overland to the Saskatchewan, but in addition nearly all her interior work—the steam engine, the boilers, the bilge pumps, and other major components—still had to be forged, assembled, and installed.

But ferocity, not of temper but of creative energy, was Sukanen's great resource against criticism, the elements, and the frustration of encroaching age. So he began the arduous task of moving the shell and soul of the *Dontianen* he had nurtured so long toward the fast-flowing waters of the Saskatchewan. He had to abandon his quarter-section homestead, the home on which he had laboured so arduously for the benefit of his supposedly waiting Minnesota family.

Planning began for the final stages of assembling the steamboat. Sukanen intended to strap together a raft at the riverside for transporting himself, a horse, and the superstructure. The raft would tow the forty-three-foot hull (complete with steam engine and boiler parts) and the nine-foot keel, which would float air-tight on its side, downriver to the point of assembly. Sukanen expected to shove off on the spring flood waters of the Saskatchewan and allow the high river current to carry him and *Dontianen* quickly into the complex lake and river system of northeast Saskatchewan and northern Manitoba. The farther downstream Sukanen rafted the deeper and stronger the water flow would be. Then, reaching the deep delta of the Nelson or the Churchill, on the threshold of Hudson Bay, he would flood his keel upright, assemble *Dontianen*'s parts, and make his way to ocean waters.

But seventeen miles and stupendous labour lay between Sukanen's creation and its planned launching. Using a system of anchored posts and winches, and his sole remaining horse, Sukanen dragged the cumbersome hull towards the northeast. With the determination of ancient engineers, he built primitive wheels, eighteen inches wide and twenty-four inches in diameter; these wheels were mounted on the underside of the keel, which lay on its side and rode behind the hull. Interminably, twenty-foot stage after twenty-foot stage, anchors driven into the ground, pulled up, moved ahead, driven again—Sukanen winched *Dontianen* toward the River. Gullies obstructed his hauling path. Sukanen fashioned wheels made by rolling willow trunks strung together with steel bands in tandem, and crossed them. Nothing could resist his determination. But after two seasons he had still only transported *Dontianen* a distance of two miles.

"I couldn't see where he'd get any place with it." Elmer Sukanen remembered the futility of his uncle's attempts to get the steamboat moved to the River. "I'll tell the truth. I don't think he ever would have made it. . . . But I helped him move part of it. It was the top of it, the captain's part where he was going to live, that we moved first. I went over with four horses and I started movin' it, and it broke. He had taken an old wagon and made it wider, but he didn't make it strong enough, so we had to rig it up and put it on a different kind of set-up. Some of the rest of it he pushed in a wheelbarrow from there to the river."

Inevitably Sukanen's health deteriorated,

physically (and according to some, mentally). Years before he had owned nine well bred horses, but with time and the Depression closing in on him, he began grinding up the last of his homestead's grain into meal for his own consumption, even the team of horses he eventually butchered and ate one by one to keep himself going. Then his last horse, the one he used to winch the barbed wire cable through pulleys that towed *Dontianen* toward the river, finally went. Of the nine thousand dollars he had accumulated for purchasing the materials for the boat, and the seven thousand dollars received for crops after construction began, every cent had been poured into the dream. In the latter years, age and malnutrition had rotted his teeth; he pulled them out with forceps he had made, and fashioned a metal mouth plate so that he could chew the tough grain meal. And still he wielded a sixteen-pound hammer all day long shaping the steel and fastening the rivets.

"He would have to swing it too, to shape that boiler up," vouched Elmer. "Half-inch steel, cold rolled steel and that chain. I can remember a bit of it when he made it; he'd pound it and round off the links. Then he heated steel for the boiler and punched rivets through it . . . hot rivets. And it was a perfect job on them rivets. They was just like a factory job. Oh yes, he knew what he was doin' all right, but he started in a bad time." "I feel sorry now," admitted a farm neighbour, "because I had the power and equipment at the time to pull his ship to the water, but I thought if I did, other neighbours would think I was crazy like him."

To some degree the problem lay with Sukanen as much as with his neighbours. No one could help him. He was too proud to take charity. And if he did take anything, he insisted upon paying back in money, supplies, or labour. Even the West family, farming the west bank of the Saskatchewan where Sukanen forged *Dontianen*'s machinery, discreetly offered help. Sukanen never took something for nothing. "I was frightened when my husband consented to Tom's working at his boat on our land . . . alone out there near him," remembered Mrs. West. "We offered him fresh eggs. . . . During the cold months we had extra old clothes. . . . I lied and said we were getting new dishes so he could have some of our clean china. . . . But we could never give him anything. He refused our help because he couldn't pay us back. He refused to fish in the river because the fish weren't his.

"He ate rotten horse meat and old wheat chaff. . . . He lived in squalor; he was so black from his stove with no chimney, that he was shiny. . . . He never drank. . . . He was soft-spoken and never laughed. He just hammered on his anvil. . . . We could hear the steady rhythm of his pounding for days on end."

Sukanen was strange, certainly, which separated him from his neighbours, but his particular sense of humour served to heighten the barrier, rather than decrease it. "I know one time his neighbours came over to my Dad," illustrated Elmer. "Tom had butchered a pig. It was a cold day and he didn't bother to skin all the hair off real good; so he left it—he hung it like that. These guys asked, 'What's the matter that you leave the hair on the pig?' 'Well,' said Tom, 'it keeps you warmer when you eat it with the hair on.' So they figured that there was something wrong and they came running to Dad right away, that they better take him away."

In the last two years Sukanen failed rapidly, and it was obvious to anyone who had known him. The man who had performed prodigies of strength, the inventor who had built the township's first thrasher, the Finn who had always given freely to anyone who asked, was rapidly approaching the point that he could not even care for himself. Health was critically important to him; at the end Sukanen had withered into a ghost of a man, barely recognizable even to his relatives who saw him "going without". The wheat he ground and ate to keep himself going was hardly edible, let alone nutritional. He sold the pieces of his now-overgrown homestead: a mower, bits of steel, utensils, for whatever food he could gather. For a time he saved what money he could spare to buy eggs from his river neighbours, the Wests, at five cents a dozen, but when his last funds ran out his diet consisted of straight wheat and nothing else. Sukanen did not own a gun, so he didn't hunt; he never fished in the Saskatchewan because he abided by the law that prohibited river fishing.

Attrition and exhaustion took their inevitable toll. The ring of Sukanen's hammer dwindled to a sporadic tapping. Both his life and his life's savings had been invested in this project; increasingly it became clear that it would not pay off. He was alone and isolated, psychologically as well as physically. "He never wanted nobody," recalled Elmer Sukanen. "If you went there he wouldn't work at all; he'd stop, 'cause he didn't want to work when anyone was watching. I used to go there once in a while with my Dad, but he'd quit working; he had to work alone . . . one thing though, he kept one day sabbath, never worked on Saturday."

The end came when Sukanen discovered that all the fruits of his years of work, the hull, the keel, and the superstructure of *Dontianen* had been stripped by vandals. This blow to his pride, this final insult, broke him. His neighbours notified the Mounties, but more to complain about his activities than to report the vandalism he had suffered. Tom Sukanen's work was halted on the technical grounds of his obstructing His Majesty's thoroughfare. Sukanen was taken to an institution hospital at North Battleford.

"Don't ever let that ship go. Don't let them tear it apart, Vic," were Tom Sukanen's last words to his last friend, Victor Markkula. Even as Sukanen spoke vandals were tearing *Dontianen* apart body and soul. Lumber was scarce on the prairies; steel was valuable; *Dontianen*, unprotected, was at the mercy of any passer-by. People with no understanding of the labour in every notch and board of the steamboat's body carried away what they needed, and smashed what they didn't want.

Tom Sukanen died on April 23, 1943, penniless and almost forgotten. The municipality offered money for his burial, and Vic Markkula, with what they say was his last forty dollars, purchased *Dontianen*'s remains from the municipality; immediately he hauled the steamboat keel, hull, and assorted machine parts to his own farm a mile or two away, where he employed the structures as granaries during World War Two. In the fifties, while planning the present Gardiner Dam-Diefenbaker Lake complex, Prairie Farm Rehabilitation Act officials hired Tom Whitely, a local engineer, to rid the South Saskatchewan's banks of debris. Whitely, only too obliging to dispose of "the work of madman Tom Sukanen", dynamited *Dontianen*'s boilers. Victor Markkula died during that era, but before his death he entreated his son Wilf, "If I die don't you let the boat be ripped up. The right man will come along to put the boat together."

Shortly before World War Two, Moon Mullin acquired a farm in the Lake Valley area, and began in earnest a hobby he had enjoyed all his life: collecting junk, artifacts, fragments, old cars, early farm equipment, rusting tools, buggies, and dozens of other odds and ends that soon spread themselves over the greater part of the farm.

Surrounded by this strange array of relics from the prairie past, Moon became more and more caught up in his interest in history. One day, for no particular reason, he recalled the story told to him more than twenty years earlier of an eccentric farmer who had built a steamboat to sail to his native Finland. "It had slipped my mind until after the Second World War," Mullin explained. "And then I got thinking about it again, about this man; I wondered how he made out. All of a sudden it just kept working on me, one thing to the other, something just driving me on; I don't know what."

Slowly Moon developed an obsessive interest in the story of Tom Sukanen. On countless occasions he travelled into the Macrorie-Dunblane territory looking for clues of the now dead and seemingly forgotten Finn. One evening Moon and a couple of companions were travelling in foul weather about a hundred miles west of Lake Valley. They barely knew the region, and by sheer chance discovered a hotel in the tiny southwestern Saskatchewan town of White Bear, where Moon and his companions sat up with the hotel keeper for conversation. The hotel keeper listened intently as Moon related all he knew about the strange story of Tom Sukanen and his steamboat dream. When Moon had finished his story, the hotelman hesitated a moment, then asked him how he regarded Sukanen—did he take him seriously, or did he see him as some sort of half-crazed madman? Finally, after Moon had assured the man of his sincere interest in Sukanen's story, the hotel keeper told him his name and of the artifact that he, Wilf Markkula, had in his possession.

Moon, who has long believed in the predetermined design of his life, was convinced that his meeting with Wilf Markkula was but further proof of the fateful role he felt himself destined to play in uncovering the mystery of Tom Sukanen. Long after Moon had gone to retrieve the steamboat from Wilf Markkula's inherited farm, he came across a letter which Tom had written to his sister sometime before he died.

"A man will follow a light," Moon recited by heart, recalling that it was a light that had drawn him to his discovering the locale of *Dontianen*. "Four times there will be men who attempt to raise this ship, and three times they will fail. And the fourth time, a man will start the raising of her. And the ship will go up. My ship will be ready, and then I shall rest. . . . "

Continued Moon, "I was the fourth man, unknowingly, to come along. Other men worked in the summertime. They had more money to move it. They could have moved it into different towns. But something always turned up to stymie them. Either the valley flooded and they couldn't take the ship out, or something happened. But with me, it just went like silk."

Moon, his obsession with *Dontianen* a dominant concern in his life, steadily worked away at making enthusiasts of everyone who would listen to him. Few could remain indifferent or sceptical in the face of his irresistible passion for Sukanen's steamboat. Among his key converts were members of the Moose Jaw Prairie Pioneer Village and Museum, originally an antique car club that blossomed into a museum when the car buffs felt some monument ought to be fashioned to commemorate the prairie pioneer spirit. And when Mullin discovered the *Dontianen* on the Markkula farm, his obsession to preserve and restore Tom Sukanen's legacy became infectious. By January, 1972, a feasibility study was complete, $700 in donations had been collected, and the museum members had gathered a *Dontianen* retrieval crew.

"We got up there about twelve miles northwest of Birsay; we had Ray Butts, a professional mover from Moose Jaw, with his diesel rigs and his big moving outfits, two of them, to take the hull and keel out of the valley," recalled Moon. "We just slipped down there, worked on her and got the ship loaded and that night she went up the hill. I can see the big diesels a-bellowing away yet. We got the boat on top and the old Finlander standing there (Torval Skelly, who had watched over the ship for years to protect it from vandals), he said, 'She's going home, boys, she's on her way.' And I can hear that yet."

Almost forgotten, all but buried by years of scrub overgrowth and a cobweb of twitch grass, nearly lost in the snowscape of a prairie winter, and only a foggy memory to those who cared to remember, Sukanen's steamboat had returned from the dead. A plain framework barely stolen from the elements that had claimed her, the awkward fragments of *Dontianen* made their way overland to their new home.

from *Fire Canoe*, by Theodore Barris

At the Wedding of the Lake

Just when I thought
this was the real dream:
    doing nothing.
A stone sleeps in Twin Willows,
Simcoe salt under the skin
and every rib grins in silence
at other people
slowly beating themselves to death
from nine to five.

A seducing sun
pries open an eyelid:
Elmhurst beach
holds its cottage children
like plums in its soft hand:
Daphne, batik bikini on bronze,
lies surrendered,
and old Mr. McIver, safely sun-glassed,
bares his rumpled chest
and savours the popsicle of the sun.

There's Maria
                sun-dancing,
whipping foam
from the Lake's shallow mouth
with tiny sticks of arms and legs;
John curls his coffee bones
to build my castles on his sand.

A blue boat springs,
divides the green seersucker skin,
froths at the prow
and wheels away towards the horizon.

And I, an old poet,
sit on this pier, a witness
to the red rabbi of sun
finally chanting the wedding psalm
for Simcoe Lake and Simcoe sky,
as a lone gull locks its wings
and glides in the amen of an arc.

Rienzi Crusz

The Cruise of *The Coot*

My father was not the only man in Saskatoon to know the frustrations and hungers of a landlocked sailor. There were a good many other expatriates from broad waters in the city, and he came to know most of them through his work, for on the library shelves was one of the finest collections of boating books extant. Some of my father's staff—who did not know a boat from a bloat—were inclined to take a jaundiced view of the nautical flavour of the annual book-purchase list, but, after all, he *was* the chief librarian.

Aaron Poole was one of those who appreciated my father's salted taste in books. Aaron was a withered and eagle-featured little man who had emigrated from the Maritime provinces some thirty years earlier and who, for twenty-nine years, had been hungering for the sound and feel of salt water under a vessel's keel. The fact that he had originally come from the interior of New Brunswick and had never actually been to sea in anything larger than a rowboat during his maritime years was not relevant to the way Aaron felt. As a Maritimer, exiled on the prairies, he believed himself to be of one blood with the famous seamen of the North Atlantic ports; and in twenty-nine years a man can remember a good many things that ought to have happened. Aaron's memory was so excellent that he could talk for hours of the times when he had sailed out of Lunenburg for the Grand Banks, first as cabin boy, then as an able-bodied seaman, then as mate, and finally as skipper of the smartest fishing schooner on the coast.

Aaron's desire to return to the sea grew as the years passed, and finally in 1926, when he was in his sixty-fifth year, he resolved his yearnings into action and began to build himself a vessel. He married off his daughters, sold his business, sent his wife to California, and got down to work at something that really mattered. He planned to sail his ship from Saskatoon to New Brunswick—and he intended to sail every inch of the way. He was of that dogged breed who will admit no obstacles—not even geographical ones like the 2,000 miles of solid land which intervened between him and his goal.

He designed his ship himself, and then turned the basement of his house on Fifth Avenue into a boat works. Almost as soon as her keel was laid, some well-meaning friend pointed out to Aaron that he would never be able to get the completed ship out of that basement—but Aaron refused to be perturbed by problems which lay so far in the future.

By the time we arrived in Saskatoon, Aaron and his boat had been a standing jest for years. Her name alone was still enough to provoke chuckles in the beer parlours, even among those who had already laughed at the same joke a hundred times. It was indicative of Aaron's singular disdain for the multitudes that he had decided to name his ship *The Coot*.

"What's the matter with *that*?" he would cry in his high-pitched and querulous voice. "Hell of a smart bird, the coot. Knows when to dive. Knows when to swim. Can't fly worth a hoot? Who the hell wants to fly a boat?"

Aaron's tongue was almost as rough as his carpentering, and that was pretty rough. He laboured over his ship with infinite effort, but with almost no knowledge and with even less skill. Nor was he a patient man—and patience is an essential virtue in a shipbuilder. It was to be expected that his vessel would be renamed by those who were privileged to see her being built. They called her *Putty Princess*.

It was appropriate enough. Few, if any, of her planks met their neighbours, except by merest chance. It was said that Blanding's Hardware—where Aaron bought his supplies—made much of its profit, during the years *The Coot* was a-building, from the sale of putty.

When my father and Aaron met, *The Coot* was as near completion as she was ever likely to get. She was twenty-four feet long, flat-bottomed, and with lines as hard and awkward as those of a harbour scow. She was hogged before she left her natal bed. She was fastened with iron screws that had begun to rust before she was even launched. The gaps and seams in her hull could swallow a gallon of putty a day, and never show a bit of it by the next morning.

Yet despite her manifold faults, she was a vessel—a ship—and the biggest ship Saskatoon had ever seen. Aaron could see no fault in her, and even my father, who was not blinded by a creator's love and who was aware of her dubious seaworthiness, refused to admit her shortcomings, because she had become a part of his dreams too.

Mother and I were expecting it, when one March day Father announced that he was taking a leave of absence from the library that coming summer, in order to help Aaron sail *The Coot* to Halifax.

Saskatoon took a keen interest in the project. Controversy as to *The Coot*'s chances for a successful journey waxed furiously among the most diversified strata of society. The Chamber of Commerce hailed the venture with the optimism common to such organizations, predicting that this was the "Trail-Blazer step that would lead to Great Fleets of Cargo Barges using Mother Saskatchewan to carry Her Children's Grain to the Markets of the World." On the other hand, the officials of the two railroads made mock of *The Coot*, refusing to accept her as a competitive threat in the lucrative grain-carrying business.

But on the whole the city was proud that Saskatoon was to become the home port for a seagoing ship. Maps showing the vessel's route were published, together with commentaries on the scenic beauties that would meet the eyes of the crew along the way. It was clear from the maps that this would be one of the most unusual voyages ever attempted, not excluding Captain Cook's circumnavigation of the globe. For, in order to reach her destination, *The Coot* would have to travel northward down the South Saskatchewan to its juncture with the north branch, then eastward into Lake Winnipeg. From there the route would turn south to leave Manitoba's inland sea for the waters of the Red River of the North, and the territories of the United States. Continuing southward down the Minnesota River to St. Paul, *The Coot* would find herself in the headwaters of the Mississippi, and on that great stream would journey to the Gulf of Mexico. The rest of the trip would be quite straightforward—a simple sail around Florida and up the Atlantic Coast to the Gulf of St. Lawrence.

Sailing time (announced by banner headlines in the local paper—MOWAT AND POOLE TO SAIL WITH MORNING TIDE) was fixed for 8 a.m. on a Saturday in mid-June and the chosen point of departure was to be the mud flat which lies near the city's major sewer outlet on the river. The actual launching had to be postponed a day, however, when the ancient and gloomy prediction that Aaron would have trouble disentombing *The Coot* from his basement was found to be a true prophecy. In the end, a bulldozer had to be hired and Aaron, with the careless disdain of the true adventurer, ordered the operator to rip out the entire east wall of his house so that *The Coot* might go free. The crowd which had gathered to see the launching, and which at first had been disappointed by the delay, went home that evening quite satisfied with this preliminary entertainment and ready for more.

Father and Aaron had reason to be thankful for the absence of an audience when they finally eased the vessel off the trailer and into the Saskatchewan. She made no pretense at all of being a surface ship. She sank at once into the bottom slime, where she lay gurgling as contentedly as an old buffalo in its favorite wallow.

They dragged her reluctantly back on shore and then they worked the whole night through under the fitful glare of gasoline lanterns. By dawn they had recaulked *The Coot* by introducing nine pounds of putty and a great number of cedar wedges into her capacious seams. They launched her again before breakfast—and this time she stayed afloat.

That Sunday morning the churches were all but deserted, and it was a gala crowd that lined the river shores to windward of the sewer. The mudbank was the scene of frantic activity. Father and Aaron dashed about shouting obscure orders in nautical parlance, and became increasingly exasperated with one another when these were misunderstood. *The Coot* waited peacefully, but there were those among us in the crowd of onlookers who felt that she hardly looked ready for her great adventure. Her deck was only partly completed. Her mast had not yet been stepped. Her rudder fittings had not arrived and the rudder hung uncertainly over the stern on pintles made of baling wire. But she was colourful, at least. In his hurry to have her ready for the launching, Aaron had not waited for the delivery of a shipment of special marine enamel, but had slapped on whatever remnants of paint he could find in the bottom of the cans that littered his workshop. The result was spectacular, but gaudy.

Both Aaron and Father had been the recipients of much well-meaning hospitality during the night, and by morning neither was really competent to deal with the technical problem of stowage. The mountain of supplies and gear which had accumulated on the mudbank would have required a whole flock of coots to carry it. Captain and mate bickered steadily, and this kept the crowd in a good humour as the hours advanced and the moment of departure seemed no nearer.

The patience of the onlookers was occasionally rewarded, as when Aaron lost control of a fifty-pound cheese—a gift from a local dairy—and it went spinning off into the flow from the sewer. The audience was entranced. Aaron danced up and down on the mud flat, shrilly ordering his mate to dive in and rescue the cheese, but the mate became openly mutinous, and the situation was only saved by the prompt action of two small boys armed with fishing poles who caught the truant cheese and steered it gently back to shore. They would not touch it with their hands, nor would anyone else, and long after *The Coot* had sailed, that cheese still sat on the mud flat, lonely and unloved.

Mutt was prominent during these proceedings. He had been signed on as ship's dog and the excitement attendant on the launching pleased him greatly. When willing hands finally pushed *The Coot* out into the stream, Mutt was poised on the foredeck, striking an attitude, and he was the first part of the deck cargo to go swimming when the overloaded vessel heeled sharply to starboard and shook herself free of her encumbrances.

The Coot came back to the mudbank once again. Mutt withdrew under the growing mountain of discarded supplies for which there was no room aboard the ship. It was not so much the sewer that had discomfited him, as it was the heartless laughter of the crowd.

Just before noon they sailed at last, and *The Coot* looked quite impressive as she swung broadside-to under the arches of the New Bridge, accompanied by a flotilla of thirty-six sodden loaves of bread that had fallen through the bottom of a cardboard container which Aaron had retrieved from the wet bilges of the boat, and had incautiously set to dry upon the canted afterdeck.

Riding my bicycle along the shore path, I accompanied them for a mile before waving farewell and then returning to the city, where, with the rest of Saskatoon, I settled down to await reports of *The Coot*'s progress.

Our newspaper had outdone itself to cover the story properly, for it had enrolled all the ferrymen along the river as special correspondents. The ferries were located every dozen miles or so. They were square scows, fitted with submerged wooden vanes that could be turned at an angle to the current so that the water pressure on them would force the ferries back and forth across the river, guided and held on their courses by steel cables that were stretched from shore to shore just below the surface. The ferrymen were mostly farmers, with little knowledge of wider waters than their own river, so the newspaper representative who visited them (himself a fugitive from a seaport town) had given each of them a careful briefing on the proper manner in which to report commercial shipping.

When, for five full days after *The Coot* left us, there was not a single report from a ferryman, we began to worry a little. Then on Friday night the operator of the first ferry below the city—some fifteen miles away—telephoned the paper in a state of great agitation to report an object—unidentifiable due to darkness—that had swept down upon him just before midnight and, after fouling the ferry cable, had vanished again to the sounds of a banjo, a howling dog, and a frightful outpouring of nautical bad language.

The mysterious object was presumed to be *The Coot*, but the reporter who was dispatched to that section of the river at dawn could find no trace of the vessel. He drove on down stream and at last encountered a Ukrainian family living high above the riverbank. The farmer could speak no English and his wife had only a little, but she did the best she could with what she had.

She admitted that she had certainly seen *something* that morning—and here she stopped and crossed herself. It had looked to her, she said, like an immense and garish coffin that could never have been intended for a mere human corpse. When she saw it first it was being hauled across a broad mud flat by—and she crossed herself again—a horse and a dog. It was accompanied, she continued, by two nude and prancing figures that might conceivably have been human, but were more likely devils. Water devils, she added after a moment's thought. No, she had not seen what had happened to the coffin. One glance had been enough, and she had hurried back into her house to say a prayer or two before the family ikon—just in case.

The reporter descended to the river and there he found the marks left by the cortege in the soft mud. There were two sets of barefoot human tracks, a deep groove left by a vessel's keel, and one set each of dog and horse prints. The tracks meandered across the bar for two miles and then vanished at the edge of a navigable stretch of water. *All* the tracks vanished—including those of the horse. The reporter returned to Saskatoon with his story, but he had a queer look in his eye when he told us what he had seen.

As to what had actually happened during those five days when *The Coot* was lost to view, my father's log tells very little. It contains only such succinct and sometimes inscrutable entries as these: *Sun. 1240 hrs. Sink. Again. Damn. . . . Sun. 2200 hrs. Putty all gone. Try mud. No good. . . . Wed. 1600 hrs. A. shot duck for din., missed, hit cow. . . . Thurs. 2330 hrs. Rud. gone west. Oh Hell! . . . Fri. 1200 hrs. Thank God for Horse.*

But the story is there nevertheless.

It was in an amiable and buoyant mood that Father and Aaron saw the last of Saskatoon. That mood remained on them for three miles during which they made reasonably good progress, being forced to make for shore—before they sank—only four times. At each of these halts it was necessary to unload *The Coot* and turn her over to drain the water out. Aaron kept insisting that this would not be necessary in the future. "She'll soon take up," he told my father. "Wait till she's been afloat awhile."

As the day drew on, the initial mood of amity

wore thin. "She'll take up all right," my father remarked bitterly as they unloaded *The Coot* for the twelfth time. "She'll take up the whole damned river before she's done—that's what she'll do!" By the time they established their night camp they had covered a total distance of six miles, and *The Coot* had lost what little putty still remained in her. Her crew slept fitfully that night.

On Monday there was little difficulty keeping the water out, since there was no water—only a continuous sand bar. It was a terrestrial day, and they hauled *The Coot* the entire two miles that they made good before sunset. The three days which followed were of a similar nature. Mutt began to get footsore from sand between his toes. Because they spent so much of their time slithering and falling in the river muck as they attempted to haul *The Coot* a little farther on her way, both Father and Aaron abandoned clothing altogether and went back to nature.

They kept making new discoveries about their vessel and her equipment, and these were almost all discouraging. They found that in the monstrous pile of stores left behind in Saskatoon had been the fuel for the stove, the ammunition for the shotgun, though not (alas for an innocent cow) for the twenty-two rifle; the axe, and, blackest of all omissions, three bottles of Jamaica rum. They found that most of their soft rations were inedible because of prolonged immersion in sewage water, and they found that their sodden blankets were in an equally unsanitary state. They found that all the labels of the canned goods had washed away, and they discovered that the two cases of gleaming, but nameless, cans which they had supposed held pork and beans actually held dog food intended for Mutt.

I do not wonder that the log had so little to say about those days. I only wonder that *The Coot* continued on her voyage at all. But continue she did, and on Thursday evening her crew was rewarded by at last reaching relatively navigable waters. It was nearly dusk by then, but neither mate nor skipper (both of whom had become grim and uncommunicative) would be the first to suggest a halt, and Mutt had no say in the matter.

They pushed *The Coot* off the final sand bar and slipped away downstream into the darkness. At midnight they fouled the ferry cable, and lost their rudder.

That loss was not so serious as it seemed to them at the time, for before dawn they were aground again—and again trudging over the mudbanks with the towropes gnawing into their bare shoulders while *The Coot* obstinately dragged her keel.

They had paused for a while in order to cook a dismal breakfast when my father, happening to glance up at the high bank, saw the horse. Inspiration came to him and he leaped to his feet, shouting with elation. He was no longer shouting when, after having hiked five miles over the burning prairie in order to find the horse's owner, and arrange for a temporary rental, he came wearily back down the banks of the river to rejoin *The Coot*. Aaron greeted him with unwonted joviality and a momentous announcement. "I've found it, Angus!" he cried, and held aloft one of the precious bottles which had been given up for lost.

It was the turning point of the journey.

By noon the amiable horse had dragged *The Coot* across the two-mile flats to open water once again. Aaron allowed the horse to wade a little way out from shore in order to float *The Coot*. He was about to halt the beast in order to untie the towrope when my father's genius renewed itself. "Why stop him now?" Father asked.

Aaron looked at his mate with growing affection, and passed the bottle. "By God, Angus," he said, "for a librarian you've got quite a brain."

So *The Coot* proceeded on her way under one horsepower and, since the river seldom was more than three feet deep, the horse experienced but little difficulty in his strange role. When, as occasionally happened, he struck a deep hole, he simply swam until he could touch bottom once again. And when the water shoaled into a new sand bar, *The Coot*'s passengers jumped ashore and helped him haul.

The use of a river horse was a brilliant piece of improvisation, and it might well have sufficed to carry the voyagers to Lake Winnipeg—where they would assuredly have drowned, had it not been for the flood.

When the rain began on Saturday afternoon, Father and Aaron took *The Coot* to shore, hauled her a little way up on the flats, covered her with a big tarpaulin, and crawled under the canvas to wait out the downpour. The horse was turned loose to scale the high banks and forage for himself, while the two men and the dog relaxed cozily in their shelter over tins of dog food and dollops of red rum.

The rain grew heavier, for it was the beginning of one of those frightening prairie phenomena—a real cloudburst. In less than three hours, three inches of water fell on the sun-hardened plains about Saskatoon and that was more than the total rainfall during the previous three months. The ground could not absorb it and the steep-sided gulches leading into the valley of the Saskatchewan began to roar angrily in spate. The river rose rapidly, growing yellow and furious as the flow increased.

The first crest of the flood reached *The Coot* at about five o'clock in the afternoon, and before her crew could emerge from their shelter, they were in mid-stream, and racing down the river at an appalling clip. Rudderless, and with only one remaining oar—for Aaron had used the other to support a tea pail over an open fire a few days earlier, and then had gone off to sit and think and

had forgotten about oar, tea, and fire—there was nothing useful that *The Coot*'s crew could do to help themselves. The rain still beat down upon them, and after a brief, stunned look at the fury of the river, they sensibly withdrew under their canvas hood, and passed the bottle.

By seven o'clock the rain had moderated to a steady drizzle, but the flood waters were still rising. In Saskatoon we who waited impatiently for news of *The Coot* were at last rewarded. The arrangements made by the newspaper began to bear fruit. Reports began arriving from ferrymen

all down the river, and these succeeded one another so swiftly that at times they were almost continuous. The telephone exchange at the newspaper office was swamped with messages like this one:

SPECIAL TO THE STAR:
SAILING VESSEL, COOT, OUTBOUND IN BALLAST FROM SASKATOON, SIGHTED AT INDIAN CROSSING AT 7:43 P.M. ON COURSE FOR HALIFAX, THAT IS IF SHE DON'T GO BUSTING INTO THE BIG ISLAND BAR AFORE SHE GITS PAST COYOTE CREEK

The Coot got by Big Island and Coyote Creek all right, for at 7:50 p.m. the watcher at Barners Ford reported that she had just passed him, accompanied by two drowned cows, also presumed to be en route for Halifax. At 8:02 she went by Indian Crossing . . . at 8:16 she sideswiped the Sinkhole Ferry . . . at 8:22 she was reported from St. Louis (Saskatchewan, not Missouri) . . . and so it went. The ferrymen tried to "speak" the speeding ship, but she gave them no reply and would not even deign to make her number. So swiftly did she pass that a hard-riding stockman who spotted her near

Duck Lake could not even draw alongside.

In the city room at the newspaper, reporters marked each new position on a large-scale map of the river, and someone with a slide rule calculated that if *The Coot* could maintain her rate of speed, she would complete her passage to Halifax in six more days.

By nine o'clock that evening the darkness of an overcast and moonless night had so obscured the river that no further reports were to be expected form the watching ferrymen. However, we presumed that on Sunday morning the observers would again pick up the trail. A number of people even drove out at dawn from Prince Albert to see *The Coot* go past the junction of the two branches of the river. They made that trip in vain. The flood passed and the river shrank back to its normal, indolent self, but no *Coot* appeared. She had vanished utterly during the black hours of the night.

All through that tense and weary Sunday we waited for news, and there was none. At last Aaron's son-in-law called on the Royal Canadian Mounted Police for help, and the famous force ordered one of its patrol aircraft up to make a search. The plane found nothing before darkness intervened on Sunday evening, but it was off again with the following dawn.

At 11 a.m. on Monday the following radio message was received in Saskatoon:

COOT LOCATED FIVE MILES NORTHWEST FENTON AND TWO MILES FROM RIVERBANK. AGROUND IN CENTRE LARGE PASTURE AND ENTIRELY SURROUNDED BY HOLSTEIN COWS. CREW APPEARS ALL WELL. ONE MAN PLAYING BANJO, ONE SUNBATHING, AND DOG CHASING CATTLE.

It was an admirable report, and indicative of the high standards of accuracy, combined with brevity, for which the force is justly famed. However, as my father later pointed out, it did not tell the entire story.

Mutt, Aaron, and Father had spent the whole of Saturday night under cover of their tarpaulin. Even after the rain stopped they did not emerge. Father said that this was because he wished to die bravely, and he could do so only by ignoring the terror and turmoil of that swollen river. Aaron said it was because they had found the second bottle of rum. Mutt, as usual, kept his peace.

When the light grew strong on Sunday morning, Father began to hope that they might yet survive and, pulling aside the canvas, thrust his head out for a look. He was stupefied by what he saw. *The Coot* had evidently managed to cover the entire distance to Lake Winnipeg in less than ten hours.

His bemused mind could find no other explanation for the apparently limitless expanse of brown water that stretched away on every side.

It was not until late afternoon, when the flood waters began to subside and the tops of poplar trees began appearing alongside *The Coot*, that the illusion was partially dispelled. It had vanished totally by Monday morning when the voyagers awoke to find their vessel resting on a broad green meadow, surrounded by a herd of curious cattle.

The crew of *The Coot* now proceeded to enjoy the happiest hours of their journey. There was no water in the boat, or under her. There was no sand or mud. The sun was warm. Aaron had found the third of the missing bottles, and Father had procured a side of home-cured bacon and five loaves of homemade bread from a nearby Doukhobor settler. Mutt was having a time with the cows. It was a fair and lovely place for storm-tossed mariners to drop their hook.

The idyll was disturbed by the appearance of the search aircraft; and shattered a few hours later by the arrival of Aaron's son-in-law as a passenger in a big red truck. A conference was called and the cruise was declared to be at an end, despite Aaron's blasphemous dissent. *The Coot* went ignominiously back to Saskatoon aboard the truck.

When he was safely within his own house, Father frankly admitted to us that he was delighted to be there, and that he had never really had much hope of seeing home again. For the rest of that summer he was content with *Concepcion*, and we spent many a happy week end on Lotus Lake, sailing her back and forth between the Anglican Church Beach and Milford's Beer Parlour.

But there is a curious postscript to the story of *The Coot*. One day in the autumn of the following year my father received a letter from Halifax. It contained nothing save a snapshot which showed a funny little craft (unmistakably *The Coot*) tied up alongside that famous Lunenburger *The Bluenose*. On the back of the snapshot was a cryptic message, scrawled large in purple ink. "Quitter!" it said.

Father would have felt badly about that, had not his friend Don Chisholm (who was assistant superintendent of one of the railroads at Saskatoon) shown him a waybill sometime earlier. It was an interesting document. It dealt with the dispatch of one flatcar, "with cargo, out of Saskatoon, bound for Halifax". And the name bestowed on that flatcar for the journey by some railway humourist was writ large on the bottom of the bill.

It was *The Cootie Carrier*.

from *The Dog Who Wouldn't Be*, by Farley Mowat

In It

The world is a boat and I'm in it
Going like hell with the breeze;
Important people are in it as well
Going with me and the breeze like hell—
It's a kind of a race and we'll win it.
Out of our way, gods, please!

The world is a game and I'm in it
For the little I have, no less;
Important people are in it for more,
They watch the wheel, I watch the door.
Who was the first to begin it?
Nobody knows, but we guess.

The world is a pond and I'm in it,
In it up to my neck;
Important people are in it too,
It's deeper than this, if we only knew;
Under we go, any minute—
A swirl, some bubbles, a fleck. . . .

George Johnston

When fortune empties her chamberpot
on your head, smile—and say, "We are
going to have a summer shower."

Sir John A. Macdonald

Blue Water Sailing

I sat on a bollard on the pier of the Humberstone lock and waited for the ship. I had a cardboard suitcase containing a pair of socks and a shirt. I had a thin dollar bill in my wallet. The year was 1941 and I was sixteen hoping for seventeen.

A shivering wind combed through the flat, brown Niagara country lying along the Welland Canal. The breath of November chilled me through my turtlenecked sweater and denim jeans. It was colder for the fellow who was with me; a young-old man with shifting eyes. He was just five days out of a federal penitentiary. He had taken a seven-year rap for safe-cracking and it was his third time around. His clothing was thin as paper so that his teeth rattled in the grey air of the morning.

We had arrived by train from Toronto, this ex-convict and myself, on travel warrants supplied by the Upper Lakes Shipping office. The war in Europe had vacuumed in hundreds of lake sailors and vessels were running tight schedules short-handed. I was fresh from the green pastures of Carden and the only ships I had seen were in picture books. My companion was at home on the lakes—when he wasn't doing time. We were both assigned to the ss *Brown Beaver*; he as a watchman, I as a deckhand.

With much anxiety I watched the approach of the little canaller steaming silently down the canal. Riding light she loomed enormously above us, grating slowly to a halt. A small boom from which dangled a length of knotted rope swung out. I gaped.

"Go on, go on," urged the ex-con impatiently, "what in hell are you waiting for?"

"Do I got to go up that?" The rope seemed to extend halfway to Jupiter.

The sailor swore. "do you want them to lower a gangway for you? CLIMB UP!"

I essayed the climb, suitcase and all, greatly assisted from topside by a burly thug in a slouch cap who wanted to know what in the obscene, pornographic name of jumped-up-Judas I was trying to do. I reached the deck panting, red and miserable, and promptly tripped over a bitt to sprawl like a fool at the feet of the thug who I was later to learn was the first mate. My companion ran up the rope like a squirrel, leaping on to the deck with my suitcase which I had managed to drop on his head.

"Watchman?" said the mate. The ex-con nodded and moved off forward and disappeared.

"And you, Barnacle Bill, must be our deckhand."

I had entertained a notion that a sailor walked around in clean white ducks, gazing on far horizons while the ship gently rocked and sea birds flew. My awakening was rude in the extreme. Four deck-hands occupied a dirty, steel-walled hole in the forward section of the ship; this hole was roughly the size of a broom closet. It contained one porthole so situated that clean air could not enter nor foul air escape. In this pen I sank onto my bunk and wished for immediate death. A red face and a mustache appeared and inquired sweetly as to my health and did I think I might condescend to attempt a little work. Red-face was the second mate.

Out on deck I was confronted by a bewildering array of wire cables, winches, hatches, masts, bitts, dead-eyes, and other wretched articles the use of which I was convinced no mortal could ever learn. The second mate set me to assist another deckhand who was trudging up and down the deck arranging wire cable. I turned to with a will, tugging with such vigour in the wrong direction that I jerked my fellow-toiler off his feet. He sat, rubbing his head in bafflement, looking helplessly at the mate.

"He's strong, Russ," said a wheelsman encouragingly.

"He's stupid," growled old Red-face.

The *Brown Beaver*'s destination was Port Arthur where she was to take on a load of wheat for the Toronto elevators. By the time we reached the locks at The Sault the second mate sourly conceded that if I didn't fall overboard before we arrived at the head of the Lakes I might conceivably make half a deckhand and be entitled to a fraction of what the shipping company was paying me.

On the long journey through the lakes and rivers I had considerable time to acquaint myself with the ship and those who sailed her. With my roommates, the other three deckhands, I could find little congeniality. They were as alien to my way of life and thinking as the manners and mores of a Zulu might have been. They seemed to have a single ambition and that was to reach port, get sodden drunk and scamper to a brothel. Now I was hardly brought up in an atmosphere of piety, but such a piggish approach to sex and drinking appeared to me then as it does now—joyless and unimaginative. Some of them, such as my watchman friend, had served time. Their penitentiary argot was harsh in my ears; I did not share their worship of safe-crackers and cop-fighters. I discovered that these fellows were not representative of the lake-sailing fraternity, but the sweepings of the city hired in emergency.

Everything was new and beautiful to me: Michipicoten lying in rock-hard fastness, dim and blue and lovely to starboard as we crossed Superior; the herring gulls hanging in the wind at mast height; the trailing streamers of smoke rolling on the horizons. . . .

At Port Arthur I drew some of my wages and was allowed to go ashore. I declined to accompany the other deckhands and set off by myself, with something of a swagger, to explore the world. Unknown to me at the time, the first mate shadowed my perambulations so unobtrusively that I was quite unaware that he was following to make sure I kept out of trouble.

Crossing windy, cold Superior on the return trip, the vessel barged into one of those tremendous autumnal gales that sweep the desert of waters, and I learned the terror of the landsman in a storm at sea. Gone were the friendly gulls and reassuring islands and the soft blue skies. The wind screamed—a harpy frenzied in the rigging; the spray rattled past like buckshot. The cramped deckhands' quarters were foul with vomit and unwashed clothing. Delighted to find that, while frightened, I was not sea-sick, I sought the deck, preferring the clean rage of the waves and winds to the putridness of the cabin. The mates allowed me sanctuary in the wheel-house; I was even allowed a thrilling five minutes at the wheel where I managed to get fifteen degrees off course in as many seconds. I admired the waves, marching like great, grey wolves, rank on rank, hurling themselves relentlessly at the bow-spar, breaking with a shudder on the ship's bow, to go flooding in angry foam about the anchor winch. I carried coffee and sandwiches along the swaying, streaming deck from the galley aft to the beleaguered mates and wheelsmen. I coughed and choked manfully on a cigar presented me by the captain. When we reached the calm of the St. Mary's River with its birch-bound shores I was almost regretful for in my ignorance I did not know that Lake Superior can swat down a ship and crew as easily as a man may slap a mosquito.

Down Lake Huron and through the belt of the St. Clair River whose summer cottages stood forlorn and empty this late in the year. Across shallow Lake St. Clair, past the towering buildings of booming Detroit city and into Lake Erie where a watery sun made a sparkle on the blue chopping waters.

I sailed for some years after that. I sailed the ocean and some seas. It was never as much fun again.

from *Fair Days Along the Talbert*, by Dennis T. Patrick Sears

The Squall

When the squall comes running down the bay,
Its waves like hounds on slanting leashes of rain
Bugling their way . . . and you're in it;
If you want more experience at this game
Pull well and slant well. Your aim
Is another helping of life. You've got to win it.

When you're caught in an eight-foot boat—seaworthy though,
You've got to turn your back, for a man rows backwards
Taking direction from the sting of rain and spray.
How odd, when you think of it, that a man rows backwards!

How odd, when you think of it, that a man rows backwards.
What experience, deduction and sophistication
There had to be before men dared row backwards
Taking direction from where they'd been
With only quick-snatched glances at where they're going.

Each strongbacked wave bucks under you, alive
Young-muscled, wanting to toss you in orbit
While whitecaps snap like violin-strings
As if to end this scene with a sudden exit.

Fearfulness is a danger. So's fearlessness.
You've got to get that mood which balances you
As if you were a bird in the builder's hand;
For the boat was built in consideration
Not only of storms . . . of gales too.

Though you might cut the waves with your prow
It'll do no good if you head straight to sea.
You've got to make a nice calculation
Of where you're going, where you want to be,
What you need, and possibility;
Remembering how you've survived many things
To get into the habit of living.

Milton Acorn

The Wreck of the Edmund Fitzgerald

The legend lives on from the Chippewa on down
of the big lake they called Gitche Gumee
the lake it is said never gives up her dead
when the skies of November turn gloomy
With a load of iron ore 26,000 tons more
than the Edmund Fitzgerald weighed empty
that good ship and true was a bone to be chewed
when the gales of November came early

The ship was the pride of the American side
comin' back from some mill in Wisconsin
as the big freighters go it was bigger than most
with a crew and good captain well seasoned
concluding some terms with a couple of steel firms
when they left fully loaded for Cleveland
and later that night when the ship's bell rang
could it be the north wind they'd bin feelin'

The wind in the wires made a tattletale sound
and a wave broke over the railing
and every man knew as the captain did too
'twas the witch of November come stealin'
the dawn came late and the breakfast had to wait
when the gales of November came slashin'
When afternoon came it was freezin' rain
in the face of a hurricane west wind

When suppertime came the old cook came on deck
sayin' "fellas it's too rough to feed ya"
at seven p.m. a main hatchway caved in
he said "fellas it's bin good to know ya"
The captain wired in he had water comin' in
and the good ship and crew was in peril
And later that night when 'is lights went out of sight
came the wreck of the Edmund Fitzgerald

Does anyone know where the love of God goes
when the waves turn the minutes to hours?
the searchers all say they'd have made Whitefish Bay
if they'd put fifteen more miles behind 'er
they might have split up or they might have capsized
they may have broke deep and took water
all that remains is the faces and the names
of the wives and the sons and the daughters

Lake Huron rolls   Superior sings
in the rooms of her ice water mansion
old Michigan steams like a young man's dreams
the islands and bays are for sportsmen
and farther below Lake Ontario
takes in what Lake Erie can send her
and the iron boats go as the mariners all know
with the gales of November remembered

In a musty old hall in Detroit they prayed
in the maritime sailors' cathedral
the church bell chimed 'til it rang twenty-nine times
for each man on the Edmund Fitzgerald
The legend lives on from the Chippewa on down
of the big lake they called Gitche Gumee
Superior they said never gives up her dead
when the gales of November come early.

Gordon Lightfoot

The Edmund Fitzgerald: It's a Year Since She Sank

ON BOARD S.S. FORT HENRY—The Henry Steinbrenner is just off our stern.

We are 460 feet and she is 730 feet with twice our horsepower, but the Steinbrenner is heading for shelter.

The U.S. Coast Guard has issued a gale warning on Lake Superior and the Steinbrenner is seeking the protection of Goulais Bay to sit out the blow.

The wreck of the Edmund Fitzgerald a year ago today, in which twenty-nine seamen died, has inspired caution on the huge and treacherous lake.

So the Steinbrenner will join four other bulk carriers anchored in the bay, waiting to sneak across Superior between storms.

But the Fort Henry is ready to test that dangerous temper again as she puts her nose into the forty miles per hour winds and the ten foot waves and heads out of Sault Ste. Marie bound for Thunder Bay 200 miles away.

The Fort Henry's skipper, Ferdinand Gagne, sixty-five, has been battling Superior's November fury for forty-eight years and so far he has always won.

The lake threw some of its best punches at Gagne and his Fort Henry a year ago because they dared venture out from Thunder Bay when so many ships wouldn't.

As the night grew long the punches became heavier and on his radio Gagne could hear the captain of the Fitzgerald, fifteen miles ahead, talking about the huge seas and the eighty-five miles per hour winds—"and he said they were pounding hell out of him, but he was still making headway."

Later that night, the Fitzgerald's captain radioed nearby ships to say he had lost some deck hatches, was listing and taking on water but still hoped to fight his way to haven in Whitefish Bay.

In the Fort Henry's log it is written the Fitzgerald disappeared from the radar screen just fifteen miles short of salvation in Whitefish Bay.

The log tells how Gagne stopped his ship that wild night to wait for sunrise so he could search for survivors from the Fitzgerald.

"All we found were a smashed lifeboat and some empty life jackets and I am glad we didn't have to pick up any bodies," Gagne said.

He has never heard Gordon Lightfoot's ballad of the Edmund Fitzgerald, though his crew men are constantly whistling the tune, and yet his words sound so much like those of the song. "Superior never gives up her dead," the captain said.

Now, as we approach the grave of the Fitzgerald and its crew, the waves are hammering at our starboard bow, the wind is screaming through the ropes and cables and we have been coated with several tons of ice since leaving Whitefish Bay.

Gagne isn't worried. "This is peanuts," the captain says.

"There have been nights out here when our stern was pointing to heaven and our nose was going straight down to hell, but still we came out of it."

Gagne smiled.

"I remember one November night when we had a 100 miles per hour wind on our port side and we had 500 tons of lumber on the deck. We went into a trough and when we came out of it there wasn't a stick of wood to be seen."

The Fort Henry is a package freighter owned by Canada Steamship Lines of Montreal. It has fifty new automobiles below deck, plus 2,600 tons of general cargo, including fifty tons of whiskey. On the open deck there are 100 tons of steel beams we picked up at the Algoma steel mill in the Soo.

Gagne calls his ship "the pony" because of its good speed, which until this spring was tops at nineteen and a half miles per hour, but damage to the propeller has cut this back to fifteen and a half miles per hour.

"I have always been on the package freighters, which is why I go through the rough weather," Gagne explained. "A bulk carrier with coal can sit and wait for good weather because there is nobody looking each day for the cargo.

"But with general cargo we are competing with trucks, trains, planes and delivery vans, so we don't sit at anchor."

The crew will tell you the "old man" loves a challenge.

"He has respect for nature, but it is going to take a helluva blow to keep him in port," said the third mate, Henry Norenz of Mississauga.

Blinding snow has now shrouded us and deckhand Roger Bienvenue of Montreal is trying to go from the forecastle 300 feet along the open deck to the dining room.

He takes only two steps. The wind catches him and he is propelled along the deck like an ice boat shooting across a frozen lake.

Fortunately, he is holding on to a lifeline—a rope hanging from a cable running the length of the ship—and he makes it aft to have his supper.

The seamen are fed as if this were to be their last meal. At this time of year it could be.

On some crossings men have gone back for meals on their hands and knees with a rope around their waist and shipmates holding the ends at both ends of the ship.

For a man to make that trip, the meal had better be worth it.

The cook, John Lewis of Toronto, hates Superior, particularly in November, because it keeps rearranging his kitchen. The lake hurls large pans, plates and mugs around the room. And scrambles the eggs before they come out of the refrigerator.

The windows in the wheelhouse are now covered with ice as the spray from the waves shoots over the bridge fifty feet above the water, but the captain is so unconcerned that he lies down on a couch in the chartroom and goes to sleep.

But there are experienced seamen on board who can't sleep.

"I'd rather be caught in a North Atlantic storm than a Superior storm any time," said Joe Shoup, of Port Colborne, an engineer in the boiler room.

"The waves on the Atlantic are further apart, and the ship rolls up and over them, but on Superior the waves are closer together and they hammer away at the boat and it is like hitting a solid wall each time," Shoup said.

And as the waves pound our starboard bow, you can hear the third mate whistling the melody of the ballad of the Edmund Fitzgerald.

Pat Brennan, *Toronto Star*, November 10, 1976

The Marine Excursion
of the Knights of Pythias

Half-past six on a July morning! The Mariposa Belle is at the wharf, decked in flags, with steam up ready to start.

Excursion day!

Half-past six on a July morning, and Lake Wissanotti lying in the sun as calm as glass. The opal colours of the morning light are shot from the surface of the water.

Out on the lake the last thin threads of the mist are clearing away like flecks of cotton wool.

The long call of the loon echoes over the lake. The air is cool and fresh. There is in it all the new life of the land of the silent pine and the moving waters. Lake Wissanotti in the morning sunlight! Don't talk to me of the Italian lakes, or the Tyrol or the Swiss Alps. Take them away. Move them somewhere else. I don't want them.

Excursion Day, at half-past six of a summer morning! With the boat all decked in flags and all the people in Mariposa on the wharf, and the band in peaked caps with big cornets tied to their bodies ready to play at any minute! I say! Don't tell me about the Carnival of Venice and the Delhi Durbar. Don't! I wouldn't look at them. I'd shut my eyes! For light and colour give me every time an excursion out of Mariposa down the lake to the Indian's Island out of sight in the morning mist. Talk of your Papal Zouaves and your Buckingham Palace Guard! I want to see the Mariposa band in uniform and the Mariposa Knights of Pythias with their aprons and their insignia and their picnic baskets and their five-cent cigars!

Half-past six in the morning, and all the crowd on the wharf and the boat due to leave in half an hour. Notice it!—in half an hour. Already she's whistled twice (at six, and at six fifteen), and at any minute now, Christie Johnson will step into the pilot house and pull the string for the warning whistle that the boat will leave in half an hour. So keep ready. Don't think of running back to Smith's Hotel for the sandwiches. Don't be fool enough to try to go up to the Greek Store, next to Netley's, and buy fruit. You'll be left behind for sure if you do. Never mind the sandwiches and the fruit! Anyway, here comes Mr. Smith himself with a huge basket of provender that would feed a factory. There must be sandwiches in that. I think I can hear them clinking. And behind Mr. Smith is the German waiter from the caff with another basket—indubitably lager beer; and behind him,

the bartender of the hotel, carrying nothing, as far as one can see. But of course if you know Mariposa you will understand that why he looks so nonchalant and empty-handed is because he has two bottles of rye whiskey under his linen duster. You know, I think, the peculiar walk of a man with two bottles of whiskey in the inside pockets of a linen coat. In Mariposa, you see, to bring beer to an excursion is quite in keeping with public opinion. But whiskey—well, one has to be a little careful.

Do I say that Mr. Smith is here? Why, everybody's here. There's Hussell, the editor of the Newspacket, wearing a blue ribbon on his coat, for the Mariposa Knights of Pythias are, by their constitution, dedicated to temperance and there's Henry Mullins, the manager of the Exchange Bank, also a Knight of Pythias, with a small flask of Pogram's Special in his hip pocket as a sort of amendment to the constitution. And there's Dean Drone, the Chaplain of the Order, with a fishing-rod (you never saw such green bass as lie among the rocks at Indian's Island), and with a trolling line in case of maskinonge, and a landing net in case of pickerel, and with his eldest daughter, Lilian Drone, in case of young men. There never was such a fisherman as the Rev. Rupert Drone.

Perhaps I ought to explain that when I speak of the excursion as being of the Knights of Pythias, the thing must not be understood in any narrow sense. In Mariposa practically everybody belongs to the Knights of Pythias just as they do to everything else. That's the great thing about the town and that's what makes it so different from the city. Everybody is in everything.

You should see them on the seventeenth of March, for example, when everybody wears a green ribbon and they're all laughing and glad—you know what the Celtic nature is—and talking about Home Rule.

On St. Andrew's Day every man in town wears a thistle and shakes hands with everybody else, and you see the fine old Scotch honesty beaming out of their eyes.

And on St. George's Day!—well, there's no heartiness like the good old English spirit, after all; why shouldn't a man feel glad that he's an Englishman?

Then on the Fourth of July there are stars and stripes flying over half the stores in town, and suddenly all the men are seen to smoke cigars, and to know all about Roosevelt and Bryan and the Philippine Islands. Then you learn for the first time that Jeff Thorpe's people came from Massachusetts and that his uncle fought at Bunker Hill (anyway Jefferson will swear it was in Dakota all right enough); and you find that George Duff has a married sister in Rochester and that her husband is all right; in fact, George was down there as recently as eight years ago. Oh, it's the most American town imaginable is Mariposa—on the Fourth of July.

But wait, just wait, if you feel anxious about the solidity of the British connection, till the twelfth of the month, when everybody is wearing an orange streamer in his coat and the Orangemen (every man in town) walk in the big procession. Allegiance! Well, perhaps you remember the address they gave to the Prince of Wales on the platform of the Mariposa station as he went through on his tour to the west. I think that pretty well settled that question.

So you will easily understand that of course everybody belongs to the Knights of Pythias and the Masons and Oddfellows, just as they all belong to the Snow Shoe Club and the Girls' Friendly Society.

And meanwhile the whistle of the steamer has again for a quarter to seven—loud and long this time, for anyone not here now is late for certain, unless he should happen to come down in the last fifteen minutes.

What a crowd upon the wharf and how they pile on to the steamer! It's a wonder that the boat can hold them all. But that's just the marvellous thing about the Mariposa Belle.

I don't know—I have never known—where the steamers like the Mariposa Belle come from. Whether they are built by Harland and Wolff of Belfast, or whether, on the other hand, they are not built by Harland and Wolff of Belfast, is more than one would like to say offhand.

The Mariposa Belle always seems to me to have some of those strange properties that distinguish Mariposa itself. I mean, her size seems to vary so. If you see her there in the winter, frozen in the ice beside the wharf with a snowdrift against the windows of the pilot house, she looks a pathetic little thing the size of a butternut. But in the summer time, especially after you've been in Mariposa for a month or two, and have paddled alongside of her in a canoe, she gets larger and taller, and with a great sweep of black sides, till you see no difference between the Mariposa Belle and the Lusitania. Each one is a big steamer and that's all you can say.

Nor do her measurements help you much. She draws about eighteen inches forward, and more than that—at least half an inch more, astern, and when she's loaded down with an excursion crowd she draws a good two inches more. And above the water—why, look at all the decks on her! There's the deck you walk on to, from the wharf, all shut in, with windows along it, and the after cabin with the long table, and above that the deck with all the chairs piled upon it, and the deck in front where the band stand round in a circle, and the pilot house is higher than that, and above the pilot house is the board with the gold name and the flag pole and the steel ropes and the flags; and fixed in somewhere on the different levels is the lunch counter where they sell the sandwiches, and the engine room, and down below the deck level, beneath the water line, is the place where the crew sleep. What with steps and stairs and passages and piles of cordwood for the engine—oh, no, I guess Harland and Wolff didn't build her. They couldn't have.

Yet even with a huge boat like the Mariposa Belle, it would be impossible for her to carry all of the crowd that you see in the boat and on the wharf. In reality, the crowd is made up of two classes—all of the people in Mariposa who are going on the excursion and all those who are not. Some come for the one reason and some for the other.

The two tellers of the Exchange Bank are both there standing side by side. But one of them—the one with the cameo pin and the long face like a horse—is going, and the other—with the other cameo pin and the face like another horse—is not. In the same way, Hussell of the Newspacket is going, but his brother, beside him, isn't. Lilian

Drone is going, but her sister can't; and so on all through the crowd.

And to think that things should look like that on the morning of a steamboat accident.

How strange life is!

To think of all these people so eager and anxious to catch the steamer, and some of them running to catch it, and so fearful that they might miss it—the morning of a steamboat accident. And the captain blowing his whistle, and warning them so severely that he would leave them behind—leave them out of the accident! And everybody crowding so eagerly to be in the accident.

Perhaps life is like that all through.

Strangest of all to think, in a case like this, of the people who were left behind, or in some way or other prevented from going, and always afterwards told of how they had escaped being on board the Mariposa Belle that day!

Some of the instances were certainly extraordinary.

Nivens, the lawyer, escaped from being there merely by the fact that he was away in the city.

Towers, the tailor, only escaped owing to the fact that, not intending to go on the excursion, he had stayed in bed till eight o'clock and so had not gone. He narrated afterwards that waking up that morning at half-past five, he had thought of the excursion and for some unaccountable reason had felt glad that he was not going.

The case of Yodel, the auctioneer, was even more inscrutable. He had been to the Oddfellows' excursion on the train the week before and to the Conservative picnic the week before that, and had decided not to go on this trip. In fact, he had not the least intention of going. He narrated afterwards how the night before someone had stopped him on the corner of Nippewa and Tecumseh Streets (he indicated the very spot) and asked: "Are you going to take in the excursion tomorrow?" and he had said, just as simply as he was talking when narrating it: "No." And ten minutes after that, at the corner of Dalhousie and Brock Streets (he offered to lead a party of verification to the precise place) somebody else had stopped him and asked: "Well, are you going on the steamer trip tomorrow?" Again he had answered: "No," apparently almost in the same tone as before.

He said afterwards that when he heard the rumour of the accident it seemed like the finger of Providence, and he fell on his knees in thankfulness.

There was the similar case of Morison (I mean the one in Glover's hardware store that married one of the Thompsons). He said afterwards that he had read so much in the papers about accidents lately—mining accidents, and aeroplanes and gasoline—that he had grown nervous. The night before his wife had asked him at supper: "No, I don't think I feel like it," and had added: "Perhaps your mother might like to go." And the next evening just at dusk, when the news ran through the town, he said the first thought that flashed through his head was: "Mrs. Thompson's on that boat."

He told this right as I say it—without the least doubt or confusion. He never for a moment imagined she was on the Lusitania or the Olympic or any other boat. He knew she was on this one. He said you could have knocked him down where he stood. But no one had. Not even when he got halfway down—on his knees, and it would have been easier still to knock him down or kick him. People do miss a lot of chances.

Still, as I say, neither Yodel nor Morison nor anyone thought about there being an accident until just after sundown when they—

Well, have you ever heard the long booming whistle of a steamboat two miles out on the lake in the dusk, and while you listen and count and wonder, seen the crimson rockets going up against the sky and then heard the fire bell ringing right there beside you in the town, and seen the people running to the town wharf?

That's what the people of Mariposa saw and felt that summer evening as they watched the Mackinaw lifeboat go plunging out into the lake with seven sweeps to a side and the foam clear to the gunwale with the lifting stroke of fourteen men!

But, dear me, I am afraid that this is no way to tell a story. I suppose the true art would have been to have said nothing about the accident till it happened. But when you write about Mariposa, or hear of it, if you know the place, it's all so vivid and real, that a thing like the contrast between the excursion crowd in the morning and the scene at night leaps into your mind and you must think of it.

But never mind about the accident—let us turn back to the morning.

The boat was due to leave at seven. There was no doubt about the hour—not only seven, but seven sharp. The notice in the Newspacket said: "The boat will leave sharp at seven"; and the advertising posters on the telegraph poles on Missinaba Street that began, "Ho, for Indian's Island!" ended up with the words: "Boat leaves at seven sharp." There was a big notice on the wharf that said: "Boat leaves sharp on time."

So at seven, right on the hour, the whistle blew loud and long, and then at seven fifteen three short peremptory blasts, and a seven thirty one quick angry call—just one—and very soon after that they cast off the last of the ropes and the Mariposa Belle sailed off in her cloud of flags, and the band of the Knights of Pythias, timing it to a nicety, broke into the "Maple Leaf for Ever!"

I suppose that all excursions when they start are much the same. Anyway, on the Mariposa Belle everybody went running up and down all over the boat with deck chairs and camp stools and baskets, and found places, splendid places to sit, and then got scared that there might be better ones and chased off again. People hunted for places out of the sun and when they got them swore that they weren't going to freeze to please anybody; and the people in the sun said that they hadn't paid fifty cents to be roasted. Others said that they hadn't paid fifty cents to get covered with cinders, and there were still others who hadn't paid fifty cents to get shaken to death with the propeller.

Still, it was all right presently. The people seemed to get sorted out into the places on the boat where they belonged. The women, the older ones, all gravitated into the cabin on the lower deck and by getting round the table with needlework, and with all the windows shut, they soon had it, as they said themselves, just like being at home.

All the young boys and the toughs and the men in the band got down on the lower deck forward, where the boat was dirtiest and where the anchor was and the coils of rope.

And upstairs on the after deck there were Lilian Drone and Miss Lawson, the high-school teacher, with a book of German poetry—Gothey I think it was—and the bank teller and the young men.

In the centre, standing beside the rail, were Dean Drone and Dr. Gallagher, looking through binocular glasses at the shore.

Up in front on the little deck forward of the pilot house was a group of the older men, Mullins and Duff and Mr. Smith in a deck chair, and beside him Mr. Golgotha Gingham, the undertaker of Mariposa, on a stool. It was part of Mr. Gingham's principles to take in an outing of this sort, a business matter, more or less—for you never know what may happen at these water parties. At any rate, he was there in a neat suit of black, not, of course, his heavier or professional suit, but a soft clinging effect as of burnt paper that combined

gaiety and decorum to a nicety.

"Yes," said Mr. Gingham, waving his black glove in a general way towards the shore, "I know the lake well, very well. I've been pretty much all over it in my time."

"Canoeing?" asked somebody.

"No," said Mr. Gingham, "not in a canoe." There seemed a peculiar and quiet meaning in his tone.

"Sailing, I suppose," said somebody else.

"No," said Mr. Gingham. "I don't understand it."

"I never knowed that you went on to the water at all, Gol," said Mr. Smith, breaking in.

"Ah, not now," explained Mr. Gingham; "it was years ago, the first summer I came to Mariposa. I was on the water practically all day. Nothing like it to give a man an appetite and keep him in shape."

"Was you camping?" asked Mr. Smith.

"We camped at night," assented the undertaker, "but we put in practically the whole day on the water. You see, we were after a party that had come up here from the city on his vacation and gone out in a sailing canoe. We were dragging. We were up every morning at sunrise, lit a fire on the beach and cooked breakfast, and then we'd light our pipes and be off with the net for a whole day. It's a great life," concluded Mr. Gingham wistfully.

"Did you get him?" asked two or three together.

There was a pause before Mr. Gingham answered.

"We did," he said "—down in the reeds past Horseshoe Point. But it was no use. He turned blue on me right away."

After which Mr. Gingham fell into such a deep reverie that the boat had steamed another half-mile down the lake before anybody broke the silence again. Talk of this sort—and after all what more suitable for a day on the water?—beguiled the way.

Down the lake, mile by mile over the calm water, steamed the Mariposa Belle. They passed Poplar Point where the high sand-banks are with all the swallows' nests in them, and Dean Drone and Dr. Gallagher looked at them alternately through the binocular glasses, and it was wonderful how plainly one could see the swallows and the banks and the shrubs—just as plainly as with the naked eye.

And a little further down they passed the Shingle Beach, and Dr. Gallagher, who knew Canadian history, said to Dean Drone that it was strange to think that Champlain had landed there with his French explorers three hundred years ago; and Dean Drone, who didn't know Canadian history, said it was stranger still to think that the hand of the Almighty had piled up the hills and rocks long before that; and Dr. Gallagher said it was wonderful how the French had found their way through such a pathless wilderness; and Dean Drone said that it was wonderful also to think that the Almighty had placed even the smallest shrub in its appointed place. Dr. Gallagher said it filled him with admiration. Dean Drone said it filled him with awe. Dr. Gallagher said he'd been full of it since he was a boy and Dean Drone said so had he.

Then a little further, as the Mariposa Belle steamed on down the lake, they passed the Old Indian Portage where the great grey rocks are; and Dr. Gallagher drew Dean Drone's attention to the place where the narrow canoe track wound up from the shore to the woods, and Dean Drone said he could see it perfectly well without the glasses.

Dr. Gallagher said that it was just here that a party of five hundred French had made their way with all their baggage and accoutrements across the rocks of the divide and down to the Great Bay. And Dean Drone said that it reminded him of Xenophon leading his ten thousand Greeks over the hill passes of Armenia down to the sea. Dr. Gallagher said that he had often wished he could have seen and spoken to Champlain, and Dean Drone said how much he regretted to have never known Xenophon.

And then after that they fell to talking of relics and traces of the past, and Dr. Gallagher said that if Dean Drone would come round to his house some night he would show him some Indian arrow heads

that he had dug up in his garden. And Dean Drone said that if Dr. Gallagher would come round to the rectory any afternoon he would show him a map of Xerxes' invasion of Greece. Only he must come some time between the Infant Class and the Mothers' Auxiliary.

So presently they both knew that they were blocked out of one another's houses for some time to come, and Dr. Gallagher walked forward and told Mr. Smith, who had never studied Greek, about Champlain crossing the rock divide.

Mr. Smith turned his head and looked at the divide for half a second and then said he had crossed a worse one up north back of the Wanapitei and that the flies were Hades—and then went on playing freezeout poker with the two juniors in Duff's bank.

So Dr. Gallagher realized that that's always the way when you try to tell people things, and that as far as gratitude and appreciation goes one might as well never read books or travel anywhere or do anything.

In fact, it was at this very moment that he made up his mind to give the arrows to the Mariposa Mechanics' Institute—they afterwards became, as you know, the Gallagher Collection. But, for the time being, the doctor was sick of them and wandered off round the boat and watched Henry Mullins showing George Duff how to make a John Collins without lemons, and finally went and sat down among the Mariposa band and wished that he hadn't come.

So the boat steamed on and the sun rose higher

and higher, and the freshness of the morning changed into the full glare of noon, and they went on to where the lake began to narrow in at its foot, just where the Indian's Island is—all grass and trees and with a log wharf running into the water. Below it the Lower Ossawippi runs out of the lake, and quite near are the rapids, and you can see down among the trees the red brick of the power house and hear the roar of the leaping water.

The Indian's Island itself is all covered with trees and tangled vines, and the water about it is so still that it's all reflected double and looks the same either way up. Then when the steamer's whistle blows as it comes into the wharf, you hear it echo among the trees of the island, and reverberate back from the shores of the lake.

The scene is all so quiet and still and unbroken, that Miss Cleghorn—the sallow girl in the telephone exchange that I spoke of—said she'd like to be buried there. But all the people were so busy getting their baskets and gathering up their things that no one had time to attend to it.

I mustn't even try to describe the landing and the boat crunching against the wooden wharf and all the people running to the same side of the deck and Christie Johnson calling out to the crowd to keep to the starboard and nobody being able to find it. Everyone who has been on a Mariposa excursion knows all about that.

Nor can I describe the day itself and the picnic under the trees. There were speeches afterwards, and Judge Pepperleigh gave such offence by bringing in Conservative politics that a man called Patriotus Canadiensis wrote and asked for some of the invaluable space of the Mariposa Times-Herald and exposed it.

I should say that there were races too, on the grass on the open side of the island, graded mostly according to ages—races for boys under thirteen and girls over nineteen and all that sort of thing. Sports are generally conducted on that plan in Mariposa. It is realized that a woman of sixty has an unfair advantage over a mere child.

Dean Drone managed the races and decided the ages and gave out the prizes; the Wesleyan minister helped, and he and the young student, who was relieving in the Presbyterian Church, held the string at the winning point.

They had to get mostly clergymen for the races because all the men had wandered off, somehow, to where they were drinking lager beer out of two kegs stuck on pine logs among the trees.

But if you've ever been on a Mariposa excursion you know all about these details anyway.

So the day wore on and presently the sun came through the trees on a slant and the steamer whistle blew with a great puff of white steam and all the people came straggling down to the wharf and pretty soon the Mariposa Belle had floated out onto the lake again and headed for the town, twenty miles away.

I suppose you have often noticed the contrast there is between an excursion on its way out in the morning and what it looks like on the way home.

In the morning everybody is so restless and animated and moves to and from all over the boat and asks questions. But coming home, as the afternoon gets later and later and the sun sinks beyond the hills, all the people seem to get so still and quiet and drowsy.

So it was with the people on the Mariposa Belle. They sat there on the benches and the deck chairs in little clusters, and listened to the regular beat of the propeller and almost dozed off asleep as they sat. Then when the sun set and the dusk drew on, it grew almost dark on the deck and so still that you could hardly tell there was anyone on board.

And if you had looked at the steamer from the shore or from one of the islands, you'd have seen the row of lights from the cabin windows shining on the water and the red glare of the burning hemlock from the funnel, and you'd have heard the soft thud of the propeller miles away over the lake.

Now and then, too, you could have heard them singing on the steamer—the voices of the girls and the men blended into unison by the distance, rising and falling in long-drawn melody:
"O—Can-a-da—O—Can-a-da."

You may talk as you will about the intoning choirs of your European cathedrals, but the sound of "O Can-a-da", borne across the waters of a silent lake at evening is good enough for those of us who know Mariposa.

I think that it was just as they were singing like this: "O—Can-a-da", that word went round that the boat was sinking.

If you have ever been in any sudden emergency on the water, you will understand the strange psychology of it—the way in which what is happening seems to become known all in a moment without a word being said. The news is transmitted from one to the other by some mysterious process.

At any rate, on the Mariposa Belle first one and then the other heard that the steamer was sinking. As far as I could ever learn the first of it was that George Duff, the bank manager, came very quietly to Dr. Gallagher and asked him if he thought that the boat was sinking. The doctor said no, that he had thought so earlier in the day but that he didn't now think that she was.

After that Duff, according to his own account, had said to Macartney, the lawyer, that the boat was sinking, and Macartney said that he doubted it very much.

Then somebody came to Judge Pepperleigh and woke him up and said that there was six inches of water in the steamer and that she was sinking. And Pepperleigh said it was perfect scandal and passed the news on to his wife and she said that they had no business to allow it and that if the steamer sank that was the last excursion she'd go on.

So the news went all round the boat and everywhere the people gathered in groups and talked about it in the angry and excited way that people have when a steamer is sinking on one of the lakes like Lake Wissanotti.

Dean Drone, of course, and some others were quieter about it, and said that one must make allowances and that naturally there were two sides to everything. But most of them wouldn't listen to reason at all. I think, perhaps, that some of them were frightened. You see, the last time but one that the steamer had sunk, there had been a man drowned and it made them nervous.

What? Hadn't I explained about the depth of Lake Wissanotti? I had taken it for granted that you knew; and in any case parts of it are deep enough, though I don't suppose in this stretch of it from the big reed beds up to within a mile of the town wharf, you could find six feet of water in it if you tried. Oh, pshaw! I was not talking about a steamer sinking in the ocean and carrying down its screaming crowds of people into the hideous depths of green water. Oh, dear me, no! That kind of thing never happens on Lake Wissanotti.

But what does happen is that the Mariposa Belle sinks every now and then, and sticks there on the bottom till they get things straightened up.

On the lakes around Mariposa, if a person arrives late anywhere and explains that the steamer sank, everybody understands the situation.

You see, when Harland and Wolff built the Mariposa Belle, they left some cracks in between the timbers that you fill up with cotton waste every Sunday. If this is not attended to, the boat sinks. In fact, it is part of the law of the province that all the steamers like the Mariposa Belle must be properly corked—I think that is the word—every season. There are inspectors who visit all the hotels in the province to see that it is done.

So you can imagine now that I've explained it a little straighter, the indignation of the people when they knew that the boat had come uncorked and that they might be stuck out there on a shoal or a mud-bank half the night.

I don't say either that there wasn't any danger; anyway, it doesn't feel very safe when you realize that the boat is settling down with every hundred yards that she goes, and you look over the side and see only the black water in the gathering night.

Safe! I'm not sure now that I come to think of it that it isn't worse than sinking in the Atlantic. After all, in the Atlantic there is wireless telegraphy, and a lot of trained sailors and stewards. But out on Lake Wissanotti—far out, so that you can only just see the lights of the town away off to the south—when the propeller comes to a stop—and you can hear the hiss of steam as they start to rake out the engine fires to prevent an explosion—and when you turn from the red glare that comes from the furnace doors as they open them, to the black dark that is gathering over the lake—and you see the men going forward to the roof of the pilot house to send up the rockets to rouse the town—safe? Safe yourself, if you like; as for me, let me once get back into Mariposa again, under the night shadow of the maple trees, and this shall be the last, last time I'll go on Lake Wissanotti.

Safe! Oh, yes! Isn't it strange how safe other people's adventures seem after they happen? But you'd have been scared, too, if you'd been there just before the steamer sank, and seen them bringing up all the women onto the top deck.

I don't see how some of the people took it so calmly; how Mr. Smith, for instance, could have gone on smoking and telling how he'd had a steamer "sink on him" on Lake Nipissing and a still bigger one, a side-wheeler, sink on him in Lake Abitibi.

Then, quite suddenly, with a quiver, down she went. You could feel the boat sink, sink—down, down—would it never get to the bottom? The water came flush up to the lower deck, and then—thank heaven—the sinking stopped and there was the Mariposa Belle safe and tight on a reed bank.

Really, it made one positively laugh! It seemed so queer and, anyway, if a man has a sort of natural courage, danger makes him laugh. Danger? pshaw! fiddlesticks! everybody scouted the idea. Why, it is just the little things like this that give zest to a day on the water.

Within half a minute they were all running round looking for sandwiches and cracking jokes and talking of making coffee over the remains of the engine fires.

I don't need to tell at length how it all happened after that.

I suppose the people on the Mariposa Belle would have had to settle down there all night or till help came from the town, but some of the men who had gone forward and were peering out into the dark said that it couldn't be more than a mile across the water to Miller's Point. You could almost see it over there to the left—some of them, I think, said "off on the port bow", because you know when you get mixed up in these marine disasters, you soon catch the atmosphere of the thing.

So pretty soon they had the davits swung out over the side and were lowering the old lifeboat from the top deck into the water.

There were men leaning out over the rail of the Mariposa Belle with lanterns that threw the light as they let her down, and the glare fell on the water and the reeds. But when they got the boat lowered, it looked such a frail, clumsy thing as one saw it from the rail above, that the cry was raised: "Women and children first!" For what was the sense, if it should turn out that the boat wouldn't even hold women and children, of trying to jam a lot of heavy men into it?

So they put in mostly women and children and the boat pushed out into the darkness so freighted down it would hardly float.

In the bow of it was the Presbyterian student who was relieving the minister, and he called out that they were in the hands of Providence. But he was crouched and ready to spring out of them at the first moment.

So the boat went and was lost in the darkness except for the lantern in the bow that you could see bobbing on the water. Then presently it came back and they sent another load, till pretty soon the decks began to thin out and everybody got impatient to be gone.

It was about the time that the third boat-load put off that Mr. Smith took a bet with Mullins for twenty-five dollars, that he'd be home in Mariposa before the people in the boats had walked round the shore.

No one knew just what he meant, but pretty soon they saw Smith disappear down below into the lowest part of the steamer with a mallet in one hand and a big bundle of marline in the other.

They might have wondered more about it, but it was just at this time that they heard the shouts from the rescue boat—the big Mackinaw lifeboat—that had put out from the town with fourteen men at the sweeps when they saw the first rockets go up.

I suppose there is always something inspiring about a rescue at sea, or on the water.

After all, the bravery of the lifeboat man is the true bravery—expended to save life, not to destroy it.

Certainly they told for months after of how the rescue boat came out to the Mariposa Belle.

I suppose that when they put her in the water the lifeboat touched it for the first time since the old Macdonald Government placed her on Lake Wissanotti.

Anyway, the water poured in at every seam. But not for a moment—even with two miles of water between them and the steamer—did the rowers pause for that.

By the time they were half-way there the water was almost up to the thwarts, but they drove her on. Panting and exhausted (for mind you, if you haven't been in a fool boat like that for years, rowing takes it out of you), the rowers stuck to their task. They threw the ballast over and chucked into the water the heavy cork jackets and lifebelts that encumbered their movements. There was no thought of turning back. They were nearer to the steamer than the shore.

"Hang to it, boys," called the crowd from the steamer's deck, and hang they did.

They were almost exhausted when they got them; men leaning from the steamer threw them ropes and one by one every man was hauled aboard just as the lifeboat sank under their feet.

Saved! by Heaven, saved by one of the smartest pieces of rescue work ever seen on the lake.

There's no use describing it; you need to see rescue work of this kind by lifeboats to understand it.

Nor were the lifeboat crew the only ones that distinguished themselves.

Boat after boat and canoe after canoe had put out from Mariposa to the help of the steamer. They got them all.

Pupkin, the other bank teller with a face like a horse, who hadn't gone on the excursion—as soon as he knew that the boat was signalling for help and that Miss Lawson was sending up rockets—rushed for a row boat, grabbed an oar (two would have hampered him)—and paddled madly out into the lake. He struck right out into the dark with the crazy skiff almost sinking beneath his feet. But they got him. They rescued him. They watched him, almost dead with exhaustion, make his way to the steamer, where he was hauled up with ropes. Saved! Saved!

They might have gone on that way half the night,

picking up the rescuers, only, at the very moment when the tenth load of people left for the shore—just as suddenly and saucily as you please, up came the Mariposa Belle from the mud bottom and floated.

Floated?

Why, of course she did. If you take a hundred and fifty people off a steamer that has sunk, and if you get a man as shrewd as Mr. Smith to plug the timber seams with mallet and marline, and if you turn ten bandsmen of the Mariposa band on to your hand pump on the bow of the lower decks—float? why, what else can she do?

Then, if you stuff in hemlock into the embers of the fire that you were raking out, till it hums and crackles under the boiler, it won't be long before you hear the propeller thud—thudding at the stern again, and before the long roar of the steam whistle echoes over to the town.

And so the Mariposa Belle, with all steam up again and with the long train of sparks careering from the funnel, is heading for the town.

But no Christie Johnson at the wheel in the pilot house this time.

"Smith! Get Smith!" is the cry.

Can he take her in? Well, now! Ask a man who has had steamers sink on him in half the lakes from Temiscaming to the Bay, if he can take her in? Ask a man who has run a York boat down the rapids of the Moose when the ice is moving, if he can grip the steering wheel of the Mariposa Belle? So there she steams safe and sound to the town wharf!

Look at the lights and the crowd! If only the federal census taker could count us now! Hear them calling and shouting back and forward from the deck to the shore! Listen! There is the rattle of the shore ropes as they get them ready, and there's the Mariposa band—actually forming in a circle on the upper deck just as she docks, and the leader with his baton—one—two—ready now—

"O CAN-A-DA!"

Stephen Leacock

The Schooner

Keen the tools, keen the eyes,
white the thought of the schooner
lined on the draughting board;
fine the stone that ground the fine blade
and skills, the many fingers
that stroked and touched it surely
til intricate delicate strong
it leans poised in the wind.

The wind that has its own ways
pushing eddying rippling invisible
in light or darkness.
Now no engineer or engine
can guide you, but
only the delicacy of touch against touch
underneath the breathing heaven.

Milton Acorn

Big Seas

In January, 1977, an oil tanker disappeared off the southern coast of Nova Scotia, only days after another split up on the rocks by Nantucket Island. The talk and concern has been of oil and pollution, but what of the sailors: What is it like to be out there, pounded by a relentless sea?

While these thoughts were in mind, we happened to overhear a neighbour, Terry Cranton of East Port L'Hebert, telling of an experience he and four other fishermen out of Lockeport, N.S., went through back in 1961. Three days adrift in a vicious storm, in a forty-five-foot Cape Island style boat off George's Bank where they'd been fishing for swordfish. Here is the story as later recorded:

We had fifty lines of gear out, fifty fathom to the line. But that was a very small amount of gear. Today they fish 500 lines. But then we hauled by hand where now they got big drums to do the hauling.

We didn't know at the time how many fish we caught that day. They just came so fast and heavy we didn't have a chance to keep track. But after we got in we had 103 fish off that many lines of trawl, which as far as I know must be a record of some kind because I never heard tell of it since. That was over two to a line which you just don't average in sword fishing.

On the average, I'd say these guys putting out 500 lines would get forty or fifty swordfish altogether and we had 103 off of fifty. So you can imagine.

We dropped on the gear about 8 o'clock in the morning. It had just started to breeze up, southeast, and the gear was all in the ball. We could only see but one or two floats. The fish had most of them under, we started hauling and all it was was swords coming up like a pin cushion.

They're all hooked in the mouth so they come up head first. As we hauled them aboard, we was cutting swords off and the fins like you normally do when you haul them in; clean 'em up and set them back, and we soon seen we couldn't handle it they was coming so fast, so after a while, we just cut fins off. We didn't bother dressing 'em and piled 'em back on the stern. After a while, it got so we couldn't even do that, they coming so fast and the way it was breezing up all the time, so we just piled them in, guts, feathers 'n all. We got the deck full and started putting them down in the hold, fins and everything, swords and the works.

We had 'em piled up I think four or five high on

the stern and we was back there on top of the fish for something, I forget what it was now, and she drove her stern under. Filled our boots, her stern went under that far.

She went under, and when she came up the dory—our lifeboat lashed on the stern—that went under too, and when she come up the sides of the dory just fell right off from the pressure inside where it was full of water.

So anyway, we filled the hold, filled the deck all we could get, and then there's like a little slaughter-house they call it where you haul your gear in, and we filled that. And that boat was full of fish! Like I say we didn't know at the time how many we had. Well the fish would average at least 100 pound, maybe 200, and there were a lot of 400 pound fish. The boat was way overloaded. But of course you go out there fishing and you don't want to fire your catch overboard.

We was quite a while getting our gear back because it was such a damned mess. We should have had it in in a couple of hours but it was such a mess, and you get sharks in among them. They come in for the bait and they get caught and get all snarled up and wound up. If I remember right, we got it back somewhere around 3 o'clock in the afternoon. It had been breezing up all along, so by then it was blowing a gale, fifty or sixty mile of wind.

Anyway we started trying to steam to windward but the wind had hauled from southeast more to the no'theast and we was pretty nigh punchin' right into it.

We kind of had a buddy boat there we went in company with, and we'd been talking back and forth. They were ten or fifteen miles from us. Then the first thing we lost contact with her. I didn't realize, perhaps not experienced, I guess. I never really been in a bad storm before.

I figured no big sweat. But then when we kept trying to get ahold of them but couldn't I said to the skipper, "I wonder what happened to them?"

The skipper, he said to me, "A storm like this, most anything could happen." He was right serious, you know? And then it come to me. That bad? And here we are in a smaller boat than they got and they might of went down.

And here we are with a hold full of fish, deck full, slaughterhouse full, and this old rotten tub. Our bilge pumps had broke down, the only thing we had to pump with was the old fashioned deck pump. The kind you sit there and rock her back and forth.

At the time he told me anything could happen, I don't think we'd steamed over two or three hours,

trying to make some headway. I guess we all realized we wasn't going to get in, but the idea was we'd try it anyway and when it gets too bad we'll stop. But it was already too bad then really.

There was a feller at the wheel, I don't know what happened, he's experienced. He's been on the water all his life, I guess, but he must have froze when he seen this big sea coming at us because the normal thing to do when you see one coming, breaking like that, is slow her down and just ride it out until after the sea's gone by. Then you can speed her up again. But he never slacked nothing. Jeez, and she kept climbing, kept climbing, kept climbing, and . . .

When you're down forward in a boat you don't have to look out the window to know when a big sea's coming. You can feel it. The boat will steady right up for some reason. It must be when she gets down in the big long swell, see, the small sea keeps her tossing and turning all the time, the big sea, when it's a long one, the boat will settle down there, right quiet and you'll hear the sea coming you know? And the first thing, "ker-thrash" and she'll fetch up. You can hear it, rushing water as it comes. It's a wonder how a boat can take it.

Well we was down forward when this big sea was coming. We knowed what was going to happen, and man alive when she let go! Everything that was laying loose, everything. Plates, covers off the stove, everything, went right up and stuck to the ceiling. The radio I was listening to, that went right up and plastered, and I lay there and watched it. For some reason or other it fascinated me. And when she hit, that radio came smack down! Hit me right on the head. She fetched up so solid. The skipper in the other bunk? Drove him down through his bunk top of the man below, wonder it hadn't killed him. Skipper was a big man, over 200 pounds.

We had to shut her down then, we just wasn't makin' no headway. Blowin' too' hard. We put the sail on to her on the stern, and just left her drift. The sail kind of holds her head to the wind so when a sea comes, instead of catching her broadside it'll kind of break on the bow. They'll take quite a breeze that way unless you're unlucky and one comes tumbling right down on top of her. We was lucky. None of them actually tumbled down on top of us or it would have cleaned her. Some pretty heavy stuff went over her, I'll tell you.

The skipper was talking on the radio to some of the other boats and told them what happened. That we couldn't steam and that the boat was way loaded with fish. And she was leaking bad. Then our set went out and we couldn't get out no more

and so of course the last report they got from us we was leaking bad, overloaded with fish and in a bad storm for a boat that size up on George's Bank. So the last report they got from us we was in trouble.

We did talk about dumping the fish and lightening her up some, but then we'd figure we'd wait for a while longer, you know? "Well, it's not being too bad now," you know. And, well, you couldn't fire your fish overboard. That's just impossible. You worked too damned hard for them.

Sometime in the night the stay wire parted on the sail. The skipper said we'd have to go back and try to get it made fast because that was what was holding the boat up in the wind and possibly the thing that was keeping us afloat.

A couple of us went back and spliced it with rope, I guess. We had to start the engines and put the boat over on the other tack so it would hold over enough that it would get the end of that wire.

Another sea broke onto her and broke the stove pipe off so we couldn't make a fire. Or water came down and filled the stove. I know we didn't have a fire. We'd open up a can of cold soup or something like that.

In every breeze there's always a slop bucket washing back and forth. I don't know why. You can never seem to get a slop bucket to stay in its place. We always used an IDO can, Irving Diesel Oil. It's their lube oil for the engines. Always cut the top out of them and use them for slop bucket.

That, everything goes into it. Dishwater, whatever, cups of tea—you dump it out and that always makes a nice mess on the floor after a sea strikes, and she goes, "Ker-sag." No one's going to clean it up in a storm, you know. It's always left to slosh back and forth. Smells nice, too.

The worst of it is you get so damned wet. Even if you don't get hit with water, down aboard a boat without any fire on, there's a dampness in everything and it'll soak right through you, your bunk. Everything gets soaked. Like when you crawl in your bunk you might as well lay down on the deck somewheres because everything's so damned wet and damp, it's miserable. And of course no heat. You're cold.

It kind of got to a point to me, I was actually hoping something would break one way or another. Either the storm would get over and we could start on our way in or else a sea would hit us and get it over with. You know, it got nerve-wracking after a while.

Every time the boat would steady up, nerves were starting to get bad, is about the size of it. Every time she'd steady up some, you'd think, "Is that a big sea coming?" You'd listen, and maybe that one would break before it got to you, or maybe break after it went past.

The "Gypsum Queen", I think it was, was on her way to Hantsport, come and layed across our bow, broke the sea off us for a while, something like three or four hours, and the wind was ninety-two miles an hour when they was laying across our bow. That was what they called "steady velocity", gusts would put it over that I would presume. It was a hurricane I guess was what it was. Late summer.

The whole time we was laying to, that was three days, we'd stand watch an hour, go back and pump for a half hour. We had to rig up lines of course, seas breaking over, you couldn't just walk back. The pump was sitting right in the middle of the deck in the open and you had to put a line on to you in case a sea struck and washed you overboard they could haul you back.

When you wasn't on watch you'd crawl into your bunk, listen to the radio. And even though I think every one of us was a bit afraid, we'd kind of get a kick out of when we'd get a report with them telling us what kind of shape we was in. They'd got the report there on the Bridgewater station that we was in serious trouble. It was kind of funny in a way.

There wasn't much talking. Most of the time everyone was to their own thoughts. I don't know if it was because they were thinking about death. That possibility. What they was leaving home. I know I certainly was. I was thinking about Kathy and what would the kids do, things like that. No one had too much to say. Just listen to the radio. That was the biggest thing. Listen to the newscast. See what they said about us.

It was the third afternoon before we could start for home. The sea had died. It was a hell of a big sea, a mountainous sea, thirty feet, I would say. It's hard to judge. But the wind had dropped out. Thirty feet sounds bad but after the wind drops out, the tops aren't breaking off so a boat the size we was in, it's no big problem because you go down in the gully and up over the top and into the gully again. When it's blowing, the tops break off and this is where you get the problem. That's what beats you to pieces.

I think what saved us was she was nothing but a glorified lobster boat. She was built low to the water, her rails were no higher than the boat I got down here now, or very little higher, and therefore with that much fish in her, the weight had her down in the water so far that everything went right over the top of her. Didn't beat her to death.

We was a day and a half getting home. Usually it would be ten hours. But we'd had to steam slow till the sea went down. We got in three or four o'clock in the morning, and I tell you old Lockeport looked some nice to me.

I'd sworn when I got in—"If this old thing gets in, I'm going to christen her. I'm going to buy a forty ouncer of rum and break it down over her rail." So the next day I went to the liquor store. But I kind of cheated. I bought a quart of wine and broke that over her rail and drank the forty ouncer.

Terry Cranton, as told to Dirk van Loon

There are a great many things that drive a man to drink, but the principal one is thirst.

Bob Edwards

Hard Times

Hard times, dearth times
Plague us every one,
Stomachs are shrunken,
Dishes are empty. . . .

Mark you there yonder?
There come the men
Dragging beautiful seals
To our homes.
Now is abundance
With us once more,
Days of feasting
To hold us together.
Know you the smell
Of pots on the boil?
And lumps of blubber
Slapped down by the side bench?
Joyfully
Greet we those
Who brought us plenty!

Traditional Eskimo Poem

Ordeal on an Ice Pan

My grandfather was my most beloved relative. Even in his last years when I knew him he was a big man with large hands and feet. Charlie Barlow loved to tell stories, to the delight of some of us and to the annoyance of Father, who believed only half of what the old man said. Father felt that these stories were only a waste of time. I joyed in them. I re-lived them. And when I went up to bed I dreamed about them. His favourite stories were about his trips to the ice and the seal fisheries and his countless brushes with death. Those were evenings before television came to our community, and after the Gerald S. Doyle news on radio I would find him in the kitchen with Grandma. He would be sitting on the couch beside her rocking chair. They would often sit together like this, saying very little, the way old people can. They never became preoccupied with what they were doing or ignored one another. Their occasional comments broke the evening stillness and they would ask me questions as I came in for the night. Quiet and solicitous to one another as they were loving to me. They moved slowly together from evening to evening into eternity. And so they shall be eternal to me, sitting together rocking in our old kitchen.

In the evening when the men came in from working, the lamp was turned on, and we would all gather in the kitchen. Grandpa would never begin a story without prompting and encouragement from his audience. I can see now that he told these stories mainly for Grandma's amusement. He was her true love even up to the cruelest extremities of old age. With a more serious man like my father, life would have been unbearable during these years. She loved Charlie's company, his comments on her knitting and her health. Best of all she loved his stories and encouraged him shamelessly. Mom often said that Grandma would be content to stay up all night listening to his tales. When he began a story the old lady became very quiet and listened to every word. She would laugh aloud at such stories as the time Grandpa and Nobby White were working in New York and went into a fruit store to buy some cherries to send home. They bought some tomatoes instead, which Nobby had never seen before. "Boy! Charlie! Them were some size of cherries," Nobby had kept saying. Years later and miles away from New York we all roared in laughter at the telling of it. Yes, even Father smiled severely.

When he told a really serious story Grandpa sat by the stove in a straight-backed chair. On winter nights especially, we retreated into our womb-like home. Outside, the drifts were piled against the house, often up as high as the windows. The snow filled in the yard, clogged the roads and covered the fences. Sometimes it would creep in under the door or in through the window casing, and we would have to stop it with a blanket or an old coat. The wind would howl around the house, tearing at the shingles. Inside, Grandpa would sit in his chair by the stove unmindful of the howling wind. He would begin a story in a low voice, relaxed and confident. His red shirt would light up in the glow from the stove. Occasionally he would poke through the grate with the iron poker to stir the coals. Half his face would be illuminated, half his moustache would be lit up as he spoke, and half his face shone red in the glow. His blue eyes were very intense as he got into his story. At first he would glance from one to the other of us, but soon he moved completely into his story and seemed to even see the men and events he described. Sometimes he would get out of his chair to re-create a scene or to describe an incident. He believed firmly in ghosts, and he could chill you to the bone and marrow without trying very hard.

One night Uncle Chalkey was visiting and Charlie was entertaining his son-in-law. Uncle Chalkey was a big fan of Charlie's and a good hand at telling a story himself. We were all in the kitchen with the stove red hot on a cold night in the winter. "I suppose you can remember nights like this to the ice, eh, Charlie?" Uncle Chalkey began with an innocent smile. Charlie smiled back and nodded in agreement, but did not begin.

"Yes," said Chalkey aloud, now agreeing with himself, "I bet there were many a night like this when poor man faced his maker on an ice pan." Charlie eyed him for a moment, suspiciously, then agreed. I glanced at Uncle Chalkey's face, but it was philosophical and innocent as he looked up at the kitchen lamp. "Nine inches of snow and sixty mile an hour winds, was on the news," Chalkey prompted. "North by northeast I make it."

"It was straight north when I fed the horse this evening," Charlie corrected him. "But I guess it hove around a point or two since then." Chalkey had tricked him into beginning. He went on, "Yes, Charlie, I guess a small change in the wind can mean a lot to a sealer on the ice."

"Yes, indeed, it can," said Charlie. "It can mean the difference between life and death. I was almost lost to the ice because of a change in the wind. We was right on the main herd of seals that year, and we ended up losing four men."

"Tell us about it," said Chalkey. We all cocked our ears as Grandpa began:

"We was nearing the heavy ice off the Horse Islands when we sighted the edge of the main herd. Every man's heart was in his throat as our vessel sailed north. We was about a half a mile from them when we struck the heavy ice, and our ship split her propeller lead. The wind that day were from the northeast and the ice were packed in to the land. Captain Swan gave the order and it was all hands over the side and after the seals. 'After the swiles!' they'd shout, and away they'd go over the ice like deer. Those men from Bonavist' Bay was hard old fellows.

"Well, we didn't hardly have to copy from pan to pan. The ice were so close together we could just walk along like along the road. The whole lot of us, 129 men, wasn't long bearing down on that herd. Like I say, it were the main herd, too. As soon as we struck the edge, those fellows laid down their flags and started killing. It was two o'clock of a beautiful Friday afternoon. There were fair winds to keep in the ice, and enough cloud to keep off the glare. There were no hint of trouble except for the occasional man slipping down between the pans and getting wet. Some of the older sealers took up their flags and gaffs and went farther on into the herd. Soon the ice was red with the blood of the young seals and our arms was wet to the elbows. I had piled eight or ten pelts by our flag when a couple of the Bonavist' men asked me if I wanted to go with them farther into the herd. I agreed and off we went. We walked on past hundreds and thousands of seals but we never stopped to kill. The baby seals lounged around fat and lazy in the sun, but we stepped over them and kept on walking. My imagination got the better of me and I guessed in my mind their worth in dollars. As far as the eye could see the ice were covered with seals.

"We walked until we was out of sight of even the farthest of our crew. We was in the very middle of the herd of whitecoats. Then, alas, for them poor animals that happened to be near by, we dropped our flags and began to kill. A fortune was there for the taking and we three dirty sealers began killing and skinning as hard as we could. Our clothes run red with blood and our faces were smeared and dripping with sweat. We didn't even notice a shift in the wind as we killed. From northwest it changed by slow degrees until it blowed west, and then again until it blowed southwest and soon it were blowing straight south. Still we killed on in a queer silence, broke only by the whelp of a young seal or the crack of a gaff across its skull.

"We killed on, never minding the wind or the

tide or Providence, or how we were going to carry so many pelts back to the ship. Our ship were far from our minds as we killed on in the dark silence of our dreams. I never even noticed that the snow begun to fall in thin wisps on the offshore wind, until one of the Bonavist' men spoke. He leaned for a moment on the end of his gaff and said, 'The wind have shifted around,' like he was cursing it. 'We got to be getting back,' he said, and this woke us from our dreams. We woke to the cold world of ice and hardship. 'The ice'll be breaking up soon with this down wind,' he said. The ice was kept in solid against the land by the northwest wind. A south wind now would surely blow it out to sea in loose pans; and us along with it.

"Already the shifting wind was stirring up. It began to crack and groan and big splits ran along and through it. Farther out we could see dark water where just before we could see seals. The ice was cracking and splitting about us but still we were slow to leave our pelts. The snow fell thicker and soon it would be dark. The ice were so loose now that we could feel the rise and fall of the sea underneath. The pan we was standing on suddenly split and half our pelts slid into the black water under the ice. We set our flags by what we had left and headed back into the rising wind. The pans was still close enough for us to copy from one to the other with our gaffs. The going was slow. Often we had to wait for a pan to come near enough for a safe jump. Soon we had to jump farther and farther to keep going, and soon we was all wet.

"After an hour we see it was hopeless. The wind were pushing the whole ice pack out to sea faster than we was moving in to land. By now it was dark and snowing and the wind were rising. The sea were choppy and we was all soaking wet. It were time to do something quick if we was to be saved. The two Bonavist' men sat down and began tearing off their clothes. I thought that they was gone crazy, and I shrank away from them. Then they begun wringing the sea water out of their clothes so as they wouldn't freeze. When I saw what they was up to, I did the same, as our ice pan popped from wave to wave with the black water all around us. Every now and then they'd be a jolt as we banged into another piece of ice, and at each bang a piece of our own pan cracked off. In the darkness I seen that our own pan were getting smaller and smaller, and so was our chances of rescue.

"By morning our pan were reduced to half its size and we was scanning the horizon for signs of help. There was none. I knew and I'm sure the other fellers knew, that we was drifting north all night, and further away from our ship. In the

morning we ate half of our few raisins, our hard bread, and wheat meal. All day we slopped along on our tiny raft of ice, which was getting smaller and smaller in them choppy seas.

"That evening the wind rose and freezing wet spray come in over our pan making our footing slippery and dangerous. We tied our ropes around the pan and used our gaff hooks to hang on to the wet ice. Behind us we could see white chunks from our pan sinking like sand in an hourglass. By the second evening, we was pushed together on a piece of wet ice no bigger than a big bed. We was riding low in the water and very soon our pan would not be able to hold up all three of us. There were no help in sight, and we knew that we couldn't survive another night. Three or four minutes in that black water and a man would be dead. The Bonavist' men was huddled together and I could hear them whispering. One of them whispered more than the other, and I guess what they was up to. 'We'll all die, anyway!,' he cried, losing patience, and not caring even if I did hear, I quietly laid hold of my gaff, for I now feared the worst. His friend wouldn't help him in his scheme, so the feller sat glumly as the water slipped in over the sides of our pan. Darkness settled on us again. Something had to happen soon. The feller sat opposite me on the pan, as we clung to our ropes. His hot eyes burned at me all the while, for he knew that all three of us would never survive the night, while two of us might. He sat staring at me. The wind and spray bet down on us, and slashed us all on the head and neck, but his eyes never moved off my face. He were trying to kill me with the hate in his glare. I sat quiet and saved my energy for what I knew was going to come. Soon one of us would be dead in the icy water, and I says to myself that it were not going to be me.

"The wind rose, and sometimes our little pan was rocked to overturning. Suddenly a blast of wind tossed our pan over a wave and down into a trough. The feller jumped to his feet and made for me across the slippery ice pan. I knew that he intended to grab, and heave me off the pan into the sea. As he came for me I struck him full in the throat with the spike of my gaff. He grabbed me with his icy hands and I twisted the spike in his throat. He fell at my feet choking and coughing blood. I grabbed for one of the ropes before I was swept off the ice, as he slid to the edge of the pan and disappeared into the black water without making a sound. He never even surfaced. He kept right on going down, down, and he might be still sinking.

"Now, with only two of us on the pan, our

chances of surviving had improved considerable. The other feller looked pretty harmless to me. When morning dawned the second day, I ate the raisins I had left. I looked at the other feller from Bonavist', and I handed him a few, but his fingers was frozen to the ropes so I had to put them in his mouth. His mind were wandering and he wasn't much good when they picked us up late that evening."

"Who picked you up?" Uncle Chalkey asked. "Your vessel?"

"Lord, no!" said Grandpa. "She were seventy miles to the southwest, with no propeller. You know, the vessel that picked us up were the 'Eva Marie', on the way to our ship with a new propeller lead. Our vessel was the 'Villa Nova', that year under Skipper Ronald Swan. We lost four men in that storm. They all had got stranded, just like we did, and were never picked up."

"No!" exclaimed Uncle Chalkey.

"No," said Grandpa in conclusion, "never picked up."

The silence of private reflection filled the kitchen as Grandpa sat back in his chair. The stove was getting low, so Mom quietly put in another junk of wood. Father had gone to bed. In my dreams that night I saw my grandfather as a young man, gaff in hand, stalk and slay a vicious saber-tooth tiger in a time and a place long past, and far, far away.

from the novel, *Goodbye Momma*, by Tom Moore

The Annual Seal Test

In the annual hysteria over killing seals, only the committed act with conviction. Only the seal-hunters and the seal-saviours behave as though second thoughts were a sign of weakness. Each side has its clubs and, after all, no one swings a club contemplatively. Between the hunters and the hunt-haters, however, there are millions of us wafflers. We have our fashionable worries. We care about food additives, insecticides, spray deodorants, non-returnable pop bottles. We seek a defensible position on the seal-hunt fight, but whose bark can we trust? Are harp seals an endangered species, or are they not?

Yes, says Canadian zoologist David Lavigne in *National Geographic*, they are. They may be reduced to "precarious levels" before 1999. No, says the Canadian government, they are not. Yes they are, says the International Fund for Animal Welfare. The federal fisheries minister and "a wealthy sealing industry" are getting ready to slaughter "the last baby seal". Nope, says government scientist Paul Brodie, harp seals are not an endangered species. Brodie has been so bold as to argue that in the Gulf of St. Lawrence, there may soon be *too many* seals. Too many for their own good, and too many for the good of the fishing industry. If there are two sides to most stories, there are four or five sides to every seal hunt story. Take, for instance, the little matter of brutality.

The government says "a firm single blow to the paper thin skull of the seal pup provides a painless and instantaneous death. . . . Also, newborn seals do not show signs of distress when they are approached by hunters." But the International Fund for Animal Welfare says "thousands of baby white coats . . . are being brutally butchered. . . . The air is full of their frightened cries. 'Me-me-me,' they cry to their mothers. Sealers move among the newborn pups, crushing their tiny skulls with wooden clubs and stripping them of their pelts. . . . Whosh, whosh, whosh—three blows of the club and the skin of the infant is a bloody pulp. . . ."

It may have been such prose that inspired Joe Clark to tell Vancouver high school kids that a Tory government, though it would not ban seal-hunting, would search for more "humane" killing methods. That is pat. There may never be a quicker way to kill a seal than a couple of good whacks on the skullbone. Thomas H. Raddall, the fine Bluenose novelist, recalls in *In My Time* (McClelland and Stewart, 1976) that, "I fired all six of my bullets into the body of a large seal, which snarled at every impact but kept on going into the sea and swam away. The dense fat closed over each bullet so that there was no hole. . . . [Later] we snatched up heavy oak staves, and killed a big one with two blows on the forehead. It was really dead, too." Raddall dismisses the furor about the brutality of Canadian seal-hunters as "nonsense".

Henry C. Rowsell, the veterinary pathologist who's executive director of the Canadian Council on Animal Care, has also witnessed seal hunts, and he says, "The death of any living creature cannot be made palatable for those visiting this type of operation, any more than can the death of domestic animals in the slaughterhouse, where the public is forbidden." By comparison with what a hog suffers on its way through an abattoir, what the "brutally butchered" seal gets is euthanasia. This is why the St. John's *Evening Telegram* denounces "publicity-seeking misfits who ignore the savage killing of other species of mammal and pick only on the most controlled and humane harvesting in the world, the seal fishery."

"Humane" is not a word I'd apply to any kind of killing but, nevertheless, the *Evening Telegram* has a point. I ask you, all you pig-eaters, devotees of French goose liver, lustful worshippers of blood-beaded sirloin, drooling lobster-lovers with your bubbling pots, and you fish gourmets who know that trout tastes best if it's still flipping when it hits the boiling *court bouillion*. I ask you now. Isn't it simply dreadful the way those brutal Newfies skull all those itty, bitty, wittle seals?

Harry Bruce

Queen of Saanich

for Jeanne Watchuk

White Leviathan with a bellyful of buses,
cars and trucks, you move into the strait,
course fixed and scheduled to the minute.
I climb the stairs with the other passengers,
but most of them turn off toward the restaurant,
where they'll spoon chowder and glance at the sea.
I walk on the sundeck alone, and find myself
drunk again on the air and last night's brandy.

The spinning radar dish by the smoke funnel
relentlessly guides us to a distant dock,
the wind hits my hair in gusts, the gulls
come and go as they please, and suddenly
a school of killer whales breaks the surface,
twenty at least, leaping in what must be joy—
air breathers committed to the oceans,
courses set for anywhere, arrival open.

Bert Almon

Killer Whale (Victoria, BC)

In perfect response to schedule,
every hour on the hour,
she is introduced in her food-free tank:
Haida, the killer whale,
named for the great Queen Charlotte tribe
first decimated,
now tamed by white civilization.

She displays a truly wonderful intelligence,
performs with almost human understanding,
wriggling a fin,
slapping the surface with a broad, black tail,
leaping,
showing a fine white hunter's teeth,
splashing the happily squealing pleasure-seekers,
her black sides sparkling in the sun.

Finally,
in response to a slight command
and a small reward of food dropped down her throat,
she rises from the water
to plant a simulated kiss on the cheek
of the man who rations out the times
when she may eat.

The crowd is astonished.
Thousands of pounds of streamlined beautiful blubber
perform like an acrobat under the hands of little man.
Leviathan,
symbol of mystery,
symbol of chaos and the unknown,
rises at a small tin whistle sound
to pay obeisance to her trainer.

The crowd is thoughtful.
They draw their breath in mute wonder
to witness modern, civilized man
at work in nature.

Robin Mathews

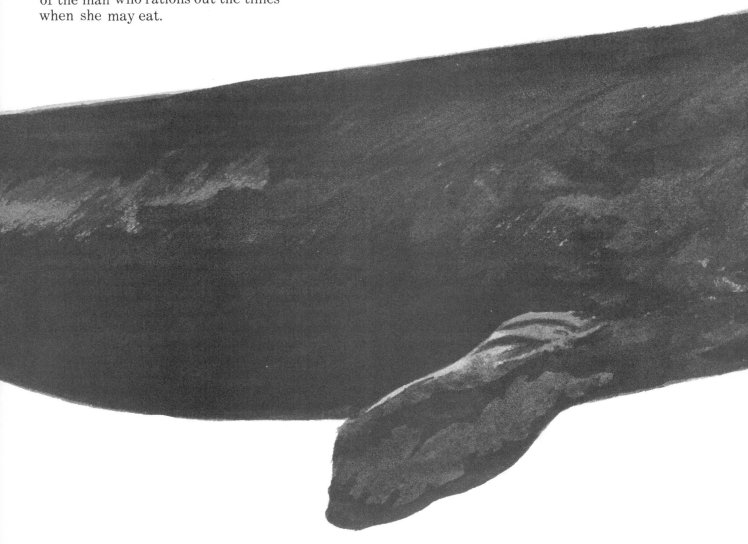

Whale Poem

Sunglare and sea pale as tears.
One long hour we watched the black whales
circling like dancers,
sliding dark backs out of water,
waving their heaved tails,
about an eyepupil-round spot
just a knife-edge
this side of the horizon.

Black whales, let me join in your dance
uncumbered by ego, my soul well anchored
in a brain bigger than I am
. . . multiplied tons of muscled flesh
roaring in organized tones of thunder
for kilometres. When I love
let me love gigantically; and when I dance
let the earth take note
as the sea takes note of you.

Milton Acorn

If Whales Could Think on Certain Happy Days

As the whale surfaced
joyously,
water spouted from his head
in great jets of praise
for the silent, awesome
mystery
he beheld between sea and sky.

Thankfulness
filled his immense body
for his sense of well-being,
his being-at-oneness
with the universe
and he thought:
"Surely the Maker of Whales
made me for a purpose."

Just then the harpoon
slammed into his side
tearing a hole in it
as wide as the sky.

Irving Layton

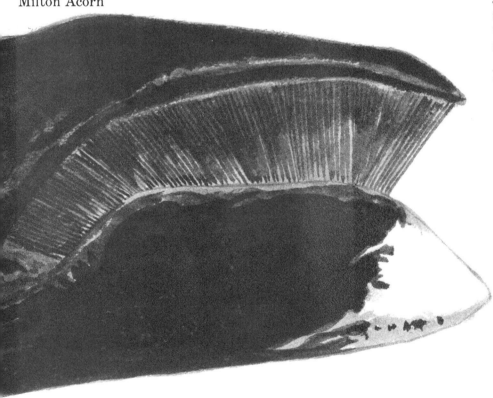

Saltwater and Tideflats

Even living on an island, one becomes very much land-bound by the habit of the road and the automobile. Travel by boat is slow and precarious, always at the control of tides and weather, and a power boat of any size is a lot to worry about when one is not using it. A car sits safely waiting, is always ready to go, and makes good time in almost any weather; ferries carry it safely and comfortably to the mainland and the roads of continent.

Yet I miss boats and the sea, the challenge of weather, the friendship of little harbours, the particular sense of freedom and reach that boats give. In the north we had nothing but boats or our feet; a seven-knot boat was fast compared to walking, it opened up a lot of new country, and it was luxuriously comfortable if it had a cabin and pilothouse. I have never liked power boats for their own sake, any more than I like automobiles or airplanes or guns for their own sake. But they are a way to wonderful things.

The thing I miss most is the sense of intimacy with the sea—not the wide sea, but the narrow coastwise seas, the channels and inlets and bays, the urgent tides, the varied, complicated shores. It still seems strange to hear a gale blowing up and not be concerned, to study the tide book only once or twice a month instead of daily, to watch the coming and going of small boats with a sense of detachment instead of intimate concern. In the north we fetched our mail by boat and all the food we bought, we went visiting by boat, often running from dawn to dark to get to a Saturday night dance, and we used boats in our work, hand-logging or fishing, even for some of our trap lines.

There were many boats, of many kinds. Once Ed and I had nothing better than a water-logged Peterborough canoe, but we crossed Johnstone Strait in it twice a week through a whole winter and never missed a mail day until the police stopped us in March. I think that was the closest intimacy with the sea I have ever had; in the trough of the wave from a good southeast gale we would see only water, angry and twisted and breaking, then the canoe would rise on the next swell, up and up until we could see the land ahead and the land behind; and the combers broke, hissing and splattering spray after us, only a foot or two from Ed's hand on the stern paddle. One had a sense of wading through the sea as through heavy wet brush, except that it was easier and more exciting; the only water we shipped was spindrift from the whitecaps and the canoe seemed able to ride anything in the world, though one could feel the thin ribs and planking bend to the heave of water under one's knees.

We worked and traveled a lot also in the *Wavey*, an open twenty-foot boat with a single-cylinder four-horse inboard. The *Wavey* was a strong little boat, very good and very dry in a sea, and handy for beachcombing logs, as one could run her nose onto a good beach, if there was not too much swell. Ed had unlimited faith in her seagoing qualities and there were times when the sea seemed as close as it seemed in the canoe. We borrowed or traveled in or visited aboard many others of all shapes and sizes, seine boats and gill-net boats and fish-packers, camp boats, mission boats, trollers, even an occasional cruiser or yacht. And once we had a float-house, a shack on a raft of logs that we towed around to our logging so that we would have some place to live beside the boat.

The boat that did most work for us, and the finest sea boat of her size I have ever known, was Buster's troller, *Kathleen. Kathleen*, big by our standards, was thirty-two feet, with cabin and pilothouse, two after hatches and fourteen heavy horsepower. Buster took her anywhere, in any weather, and we were often intimate with the sea when the green swells battered themselves against the pilothouse windows. One night we ran her bow onto a shelving shingle beach in a storm and I went overboard with a heavy pry to hold her straight until the engine dragged her off. I grabbed for the bow as she slid away, thinking to pull myself aboard; but my thigh boots were full of water and I hadn't the strength, so I clung there, hollering for Buster and wondering whether to cut loose and try to swim back to shore. Buster supposed I had stayed ashore when the boat came away and heard nothing over the sound of the storm and the motor; but some seaman's sense brought him forward when *Kathleen* was clear and he found me there and pulled me aboard. I still wonder sometimes, in fact I'm wondering now, if I could have made the swim in thigh boots and a heavy mackinaw.

We were in trouble at other times with *Kathleen*. Once a plug came loose in one of the live boxes when we had five tons of freight aboard. We tried to swing her towards shore, but a steering chain broke and there was nothing to do but leave the freight out on deck and bail until we could find the plug and drive it back. And once Buster took a big wave over the stern off Cape Mudge; it smashed in the pilothouse door and killed the engine and left him bailing for six long hours until the tide turned and the sea quieted and he could get started again. But I remember her best when we were working with her, towing a boom of logs through the calm night and into the next dawn, searching along a beach for stranded logs, swinging in to drop anchor and go ashore with the tools in the dinghy.

Buster was very proud of *Kathleen*'s dinghy. It was eight feet long, clincher built, weighing forty-five pounds—an easy single-handed heave brought it clear up on deck or slid it overside. But it was a tough little boat to launch from the beach in a swell; I tried three times one day and each time a wave caught it, pitched it back like a piece of cork, and left me sitting in water up to my neck. The next time we made it, but left the tools ashore for a calmer day. I saw *Kathleen* many times from water level, because Buster used to urge me to ride the logs out to the boat when he towed them off. We would jack one up into a good position, Buster would run in with *Kathleen* and throw me a tow line. Then I would drive a log and set three or four turns of the line around the log; Buster would take a run with *Kathleen*, the log would roll and splash into the water, still rolling, and I would jump wildly for it. Sometimes it all went well; I stayed on and Buster slowed *Kathleen* and took in the towline, I stepped aboard, and we were ready to go again. Far too often for my self-respect I pitched into the water and had to steady the log and climb on to it again somehow, while Buster anxiously circled *Kathleen* to the rescue.

Now I travel more or less sedately in the police boat from time to time, to try a case somewhere out among the islands. Or we charter sometimes cruisers, sometimes fishing boats, to go out and hunt ducks and geese and brant. I am still happy with it and the many pleasures and concerns of the sea come back at once, as though I had never been away from them.

Roderick Haig-Brown

Thoughts Under Water

Modern inventions in fly-fishing tackle do not impress me greatly. Glass rods have power and lightness and cheapness, as well as a measure of durability, but there is really nothing they will do that cane rods cannot do just as well. Synthetic lines have their virtues, but silk lines in competent hands will do just about everything that can be claimed for synthetic lines and they left us free from the awful modern complication of weights, and comparatively free from the wretched decision of which lines to take along for any given set of conditions. Synthetic leaders are stronger for their thickness, but in most respects they are inferior to silkworm gut and it is a pity that their cheapness, convenience and availability persuade us to use them. Good fly reels have needed no improvement in my fishing lifetime. And nothing much has been added to the art of fly-typing, dry or wet, since hair wings first became popular.

But the invention of the wet suit, mask, snorkel and flippers is in an entirely different category. It has added an entirely new dimension to my life and I am eternally grateful to the men who developed the gear and techniques to their present state of perfection. For the first time a fisherman can go through the surface film in comfort and stay there just as long as he likes. For the first time a man can experience some of the sensations of a fish and can know the bottom of a river as well as he knows its surface. Just a few years ago nothing of the sort was possible, even for the hard-hat diver who walked the bottom at the end of his air line. Today anyone can do it.

Even when I decided to invest in the necessary equipment I had only the slightest idea of what to expect from it. I supposed I should find cold water somewhat less cold, that I should be able to poke about briefly here and there in the quieter parts of the streams and see fish better than I had seen them before. Instead I found myself immediately transported from the world of air to the world of water and at least as comfortable in the new world as the old. True, there were occasional gaggings and sputterings as I learned the rhythms of breathing through a snorkel, moments of doubt as I experimented to find just how much weight I must carry on my belt to give me approximately neutral buoyancy, some aching of the knees and leg muscles as I accustomed myself to the use of flippers. But these were trivial discomforts that serve only to emphasize the miraculous thing that was happening to me.

Many years ago, when the children were small and learning to swim, I built a wing dam out into the river by piling up rocks and boulders. It was a laborious process, as we had no other machinery than a horse and a set of blocks, but the result has stood firmly against the freshets of more than twenty years and now the controlled river runs much of the time at a height that just breaks over the dam in three or four bubbling runs, while the main force of the current swings past the point in a formidable sweep of power.

This was a natural place to learn familiarity with the simple mechanics of swimming and diving, but to my delight it was also a place full of life and beauty. The moment one's masked face is under the surface film, the everyday world is lost. The body in its buoyant suit stretches out in the enclosing water; flippers, given depth by the weight belt, have a slow, easy power that is much like walking. The water, enclosing, is all-supporting. The body has no weight, only an easy fluidity of motion. One moves lazily, because the gain of violent movement is slight and exhausting. It is easy to rest, more completely than in the softest of beds, by relaxing every muscle and accepting the water's infinitely gentle support.

As I move up from the tail of the pool below the dam there are always under-yearling cohos and steelheads about me, lively little fish that care nothing at all for my strange shape and stranger movements. They will swim within inches of my outstretched hands and sometimes between them. Occasionally one seems to swim deliberately and curiously towards the face mask, perhaps attracted by the movement of the wondering eyes behind it.

A little farther up the pool, along the edge of the current that comes past the point of the dam, the fish are yearling steelheads, four, five and six inches long. During the early part of the season, in the more sheltered water, there is a scattering of greenbacked king salmon fingerlings, most of them within weeks of going to sea. I pass among them gently, looking down at the bullheads and caddis larvae on the sandy bottom. To my right the current races unevenly over a bottom of round boulders that disappears in blue-green distance. If I glance up I can see the rippled, bouncing surface.

At the head of the pool, in the plunging bubbles of the runs and among the big boulders of the dam, there are larger fish, seven- and eight-inch rainbows, probably pre-migrant steelheads in their second or even third year, and two or three sizable trout between twelve and fifteen inches. This is where I hold, schooled with the fish, my hands moving occasionally like pectoral fins, flippers

moving when they must to keep me in station. The smaller fish seem to accept me completely. The larger ones hold station or continue their affairs, but they are aware of me and will move away if I reach a hand within eighteen inches or a foot of them.

For several weeks this summer there were two large fish that held station close under the dam, both rainbows, each about fourteen inches long. The larger was a firm, bright, deep fish, the other a slender fish with a marked red stripe along its side. They seemed entirely different types and I suspected that the slender one was a resident river rainbow while the other may have been an estuary fish that had moved up. When the river was high they often moved back and forth between the runs that broke over the dam and sometimes held station facing downstream among the larger boulders, searching the drift that came back in the underwater eddies. Even in this position they would allow me to approach very closely provided I did not seem to block off the obvious way of escape from among the rocks. The moment I showed any sign of doing so they darted past me into the open water of the pool.

Needless to say I became fond of these fish and I like to think that they got used to me, or at least became satisfied that I was harmless. As a fly-fisherman I was impressed first of all by their readiness to move about in response to comparatively slight changes in river height and even, it seemed, to range back and forth between preferred stations in search of the best one. But they did hold, and hold very firmly at times, especially in the violent turbulence just off the point of the dam. The boulders here are very large, some of them as much as three or four feet in diameter, and the current has dug a deep rock-floored race just beyond them. Bubbles break down in intermittent showers from the build-up behind the shoulder of the dam and nothing is constant except movement; but one can find a handhold on one or other of the big boulders readily enough and cling there, tossed and tumbled like the fish themselves.

For they are tossed and tumbled, lifted and dropped, by the swirls and eddies and surges. I had imagined trout holding in such a place well down on the bottom, taking shelter among the round rocks and darting out to intercept drifting feed. These fish did not. They held just above the tops of the boulders and accepted the current with their whole bodies. They held station with the power of their bodies, even as I held with my hand's grip, and they were twisted and turned and battered by it, even as I was. But they were able to move in it as I was not—up or down or sideways with little effort, intercepting and swallowing things too fast and small for me to see. How long a fish may hold in such an active station I do not know, but I have watched them for fifteen or twenty minutes at a stretch and when some clumsy motion of mine displaced them, they returned almost immediately.

While these fish can move with a flick of their tails from the strong current into the easier flow behind the dam, I doubt if it would normally occur to me to search such heavy water with a fly. From now on, of course, I shall do so. A deeply drifted fly would be the most likely to bring a faithful response, but drifting it accurately would not be easy. A wet fly hung from above would be the next choice, but I doubt if the chances of hooking the fish securely would be better than fifty per cent. A good big dry-fly cast upstream would almost certainly produce a rise, but again the chances of hooking the fish would not be good, since he would be responding through some four feet of very fast water.

Underwater thoughts such as these are purely academic and altogether kindly towards the fish. I have promised myself to kill nothing under water and to disturb whatever lives there as little as possible. It wouldn't seem right to watch the fish behind the dam through an afternoon and go out and catch them in the evening. But academic knowledge has its uses and I have no reservations about applying what generalizations I learn under water when I am back above the surface film again and fishing in the ordinary way.

When I ask myself why I have taken to the mask and snorkel with such enthusiasm after all these years, the first answer that comes to my mind is: curiosity. To that I must add: love of the fish and love of the water. But I had no idea it would be so beautiful. I had no idea I would see so much or so clearly. I had no idea it would be possible to move about a great rushing river so freely and easily or that the human body could adapt so completely and readily to flowing water. We have disadvantages, of course. Through the best of masks our forward-seeing eyes give only restricted vision. We cannot begin to make headway against anything more than the most moderate current. It is rarely possible to approach a fish unseen, so one is always left wondering whether its behaviour is entirely natural. But in spite of all this one can see so much and so much more thoroughly than is ever possible from above the surface that it is fair to say one becomes an authentic part of the underwater world.

The underwater world is friendly and confiding, which is why I do not want to bring death or unnecessary disturbance into it. Fry and fingerlings and yearlings seem to have no fear at all of the hovering figure in the black suit. Larger trout are wary, especially when they are freshly up from salt water, but they accept the intrusion so long as the approach is not too close and in time they seem to understand there is no danger. The salmon are very nervous when they first come in, decreasingly so as they approach maturity. Rivers, even big ones, are narrow and shallow places compared to the ocean depths and a diver in a wet suit must look like nothing but danger.

Sometimes I wonder if a suit of some other colour than black might disturb the fish less; the effect of black under water is a shadowy gray, a good deal like that of a seal or a predatory fish. But it seems to me that the attention of the fish is always for the face mask and the eyes behind it—especially, I imagine, for the eyes. They watch me, eye to eye, and scatter from my slowest approach, but if I turn and look back I see them as often as not regrouped close under my moving flippers; often, too, a fish will hold almost indefinitely while one is watching him from a safe distance, then disappear immediately the eyes are turned away.

"What are you doing down there?" small boys ask me along the river. "Getting lures off the bottom?"

"No," I answer. "Looking for fish."

"Find any?"

"Sure, lots."

"What do you do when you find them?"

"Just watch them."

This final answer rarely satisfies and I could, I suppose, give others equally truthful—that I am checking on the runs, looking over the river bottom, testing the set of the currents, trying to solve a lifetime of mysteries. Some of my friends urge me to get an underwater camera and no doubt it would be nice at times to have one. But I would rather be without encumbrances and complications beyond the essential ones, at least while everything is so fresh and new. I want to be free to watch and think and feel and perhaps to come a little closer to understanding what the underwater world is really like.

Roderick Haig-Brown

Roderick Haig-Brown first knew the chalk streams of England in his boyhood. In his early twenties he earned his living partly by fishing the coastal waters of British Columbia. By his late twenties he had settled by the Campbell River on Vancouver Island and just as that river flowed past his study window it also flowed through much of his writing. He fished it, studied it, canoed on it, dove under its surface and had the sound of it constantly with him.

He usually wrote in essay form, mainly about fish and fishing; but there are also several novels on such subjects as fish, animals, West Coast Indians and logging. Besides his writing, Haig-Brown devoted much of his time in later life to his work as a judge (although he never went beyond high school) and as a conservationist, particularly of the water and the fish that live in it. He worked constantly to preserve the Fraser River and its salmon and served on the International Pacific Salmon Commission. At the time of his death in 1976 he had just completed a book on sport fishing in Canada.

Ripple Rock

Don went out to help his uncle milk before breakfast the next morning. Coming back from the barn, Joe Morgan said, "If things get tough while you're up there, don't ever forget you've got lots of friends."

"I know that, Uncle Joe," Don said simply. "But I don't reckon a man should call on his friends without he has to."

"That's up to you, but don't ever forget we're here. Mind if I say something, Don? Kind of advice, the way I used to talk when you were younger?"

Don stopped in the pathway, holding the two full pails. "Heck, no, Uncle Joe. Why should I?"

"Well, it's this. Last winter, when you and Tubby went trapping together, you had a whole lot to look out for. But they was all nature things —rivers and weather and the woods and animals. Those things can do you dirt if you don't handle yourself right, but they're still pretty simple alongside men. This fishing'll bring you up against men, as well as all the rest. Most of 'em will be for you, a few will likely be against you. It's the ones that are for you that can do you the worst harm, if you ain't watching." Joe paused and set down the pails he was carrying. He was thinking hard, searching for words to say what he wanted in a way that Don would accept, and that was unusual for Joe Morgan, because he could nearly always say what he had to say easily and well. "Fishermen are the same as other men mostly," he went on at last. "There's good and bad in them. Most of them are steady and hard workers. But there's a lot of them are kind of heedless, too easygoing and good-natured for their own good. That's where a young fellow has to watch himself. Easygoing habits are easy to get into and there doesn't seem much harm in them. There isn't, so far as that goes, except for a man's own self."

He paused again and Don said, "I think I know what you mean all right, Uncle Joe, and I'll watch it."

"There's something else," Joe said. "Liquor. There's no harm in that either, if it's used right. But don't be in any hurry to start using it. The law says under twenty-one is too young for a man to start drinking and the law's dead right for once. There's never any reason for a young fellow to start drinking then, except he thinks it's a smart thing to do."

Don asked quietly, "When did you have your first drink, Uncle Joe?"

Joe's face relaxed and he laughed aloud. "Before

I was twenty-one, I guess," he admitted. Then he was serious again. "But it wasn't more'n a glass of beer, at that. I was twenty-five and over before I tasted hard liquor and I reckon I still haven't drunk up as much money as many a man does in one year. I ain't worrying none about you," he added. "It's just I figure I ought to say what your dad would be saying to you if he was still alive."

Joe Morgan picked up his milk pails and they walked together back to the house.

Tubby arrived on time and they got a start from the river a good hour before high water slack. "Should make the Narrows fine," Tubby said.

Don did not answer. He was feeling the wrench of leaving Starbuck River, the Morgan farm, and the woods and logging works of the valley. It was a feeling that would be gone as soon as they were well on their way, he knew, but it was strong in him for the moment, not less strong because of the ten-dollar bill that Aunt Maud had slipped almost furtively into his hand as she said good-by. "It won't be much help," she had said. "But it'll buy something good to eat on the way."

They passed the bend by the big spruce tree and came in sight of salt water. A trolling boat was anchored just inside the mouth of the river. "Who's that?" Don asked.

"Don't know," Tubby said. "Never saw her before."

They studied the boat with fishermen's eagerness as they came closer to her. She was big, maybe thirty-eight feet, Tubby judged, with forty-foot poles and power gurdies to handle three lines on each side. A man came out on deck, dumped something overboard, and stood watching the *Mallard* as she came closer. He was a tall man, rangily built, yet heavy above the waistline, with dull red hair and a big handsome face. Probably about thirty years old, Tubby thought, and he weighs two hundred and ten or better. He read the name on the bow of the boat—*Falaise*.

The man waved as they came close and Don slowed his engine right down and kicked the clutch out. The man asked, "How's fishing, chums?"

"Not so hot," Tubby told him.

"What's your hurry then? Better tie up and have a cup of coffee."

Tubby started to say something about catching the tide in the Narrows, but Don kicked in the clutch and speeded up to circle and come back against the current. Tubby went up to the bow and picked up the line. "Your anchor hold us both in the current?" he asked.

"Sure thing. When Red Holiday puts an anchor out, it'll hold the *Queen Mary* in a gale."

They spoke their names and shook hands. "Partner's still asleep down below," Red told them. "Name of Moore, Tom Moore. They don't come any better. Where you fellows heading?"

"North," Don said. "Pendennis Island, I guess." He was admiring the bright steel lines on the gurdey reels and the whole workmanlike rigging of the boat. "Where're you heading?"

"I was figuring on the West Coast. But come to think of it, Tom and I are about due for a change. Why don't we go up to the Island, too? Haven't been up there since before the war. A guy can get into a rut if he goes the same place all the time."

Tom Moore came up from below, a slender, dark man with black eyes and nervous hands. "Meet Don and Tubby, Tom," Holiday said. "We're all going north together."

"We better get going pretty soon," Tubby said. "Or we'll miss the slack water in the Narrows."

"Couldn't matter less," Red told him. "If we stop for a mug-up now, we'll get us a good boost through on the start of the ebb."

"And be bucking the flood by midafternoon," Don said.

"What's your rush?" Red asked. "We'll tie up as soon as she starts to get tough and run on the ebb again after dark."

They did it Red's way and came in sight of the Narrows a full hour after slack water. Don watched the swells and folds of current on the oily-smooth water and felt the thrill he always felt in passing that place, where the powerful tides of the east coast of Vancouver Island crowd into a bare half mile of channel. He was standing with Red in the roomy pilothouse of the *Falaise*. The *Mallard* was tied alongside and Tubby and Tom Moore were both in their bunks. Don watched the steep shore line slip by faster and faster as the tide caught them. From time to time he glanced back at Red's intent face.

"Won't be very strong today," Red said without taking his eyes off the water. "We're still early for the best of it. Nothing much this side of Ripple Rock on the ebb anyway. We'll stay tied together." He spun the wheel to meet a sudden throw of unseen current and glanced quickly toward the high point south of Menzies Bay on the Vancouver Island shore. "You can run in fairly close to Race Point on the ebb. Keep right away from the darn place on the flood, though. I've seen boils of water there'd make you think the whole ocean's coming up at you from the bottom. Bad whirls, too. I'll take you up the steamer course this trip." Don watched Race Point slide by and saw Red ease the *Falaise*'s bow over toward Maud Island Light on the other shore. He felt the same elation he felt in running his canoe through a stretch of fast water in the Starbuck—a sense of power and control, yet an intense awareness of greater power, beyond control, all about him.

Red glanced toward him and grinned. "Makes a guy think, don't it? Makes you feel kind of small, but kind of good, too. There's lots worse places for a small boat than the old Seymours, but a guy gets a different feeling out of them."

It was a sparkling day, without a breath of wind, and no other boats were passing the Narrows at that time. Don watched the broken water beyond Ripple Rock. "The old Rock really kicks it up," he said. "But I should think a small boat'd handle it if a man didn't do anything crazy."

"He'd have to be crazy to go in there," Red said. "But you're probably right, at that. If a boat's a good model and you give her a fair chance, she'll handle almost anything in the way of rough water." They were passing Maud Island within two or three hundred feet, passing the shore at twice the *Falaise*'s top speed. The current swirled and pulled at them and Red moved the wheel constantly to meet the jolting shocks and twists. Suddenly the overfall of meeting currents was breaking white all about them in short, sharp little waves. The bow of the *Falaise* dipped, wrenched sharply over and Red pulled her back. Something slid across the floor of the cabin below them. "Better stand to your own boat till we see how it is," Red told Don. "I wouldn't want one of those lines to break. Leave the wheel alone though."

Don went out of the pilothouse and dropped down to the deck of the *Mallard*. Tubby came up from below. "Gee," he said. "We're taking an awful pitching around." Then he looked out at the broken water. "H'm," he added, "not really bad, considering."

"No," said Don. "That guy's good. And he likes it." The edge of a whirl caught them and both boats heeled far over, righted, hit another whirl and heeled again. "Couple more like that," Tubby said judicially, "and something's liable to break loose. I better look." He went below again.

Don leaned against his pilothouse, watching the break over Ripple Rock behind them. The water was still rough, but they were through and away for the north. It would be three months or more before he saw the Narrows again and came back to his own part of the coast.

from the novel *Saltwater Summer*,
by Roderick Haig-Brown

The Death of the Salmon

Do they all die after spawning? Yes. Even the pink salmon after a life of less than two years? Even the great and powerful king salmon? Even the jacks, the precocious males that come back after only one short year in the sea? Yes, they all die. Not a single one of all the hosts upon hosts that come in from the sea lives to spawn a second time.

It is natural for a man to resent this, I suppose, to feel that it is wasteful and shocking, in some way unnatural. Many years ago, when I first came to the rivers of the Pacific salmon, I refused to believe it. After all, some steelhead and Atlantic salmon live to spawn a second time, even a third and fourth and fifth time. One sees them, bright and clean and strong again, in the rivers after spawning and knows that the power of recovery is in them. For years I searched among Pacific salmon for some sign of recovery, for even one fish that seemed to have renewed its grip on life. I did not find it. I have seen chum salmon back in salt water but invariably they were pathetic, worn-out creatures still in the immediate process of dying. Now I have lived so long with this fact of collective, simultaneous death that I no longer resent or question it. Instead I find it fitting and beautiful, certainly useful in some way or ways that are not entirely clear and a yearly occasion of high drama. I am still curious about the manner and meaning of it, but I do not question that it has manner and meaning.

In many ways the short-run rivers are best for watching the spawning salmon. When fish have run several miles upriver from the sea, one expects them to be battered and scarred and weary. But in the short-run rivers it is clear that any changes are in them and of them. One sees everything, from the first arrivals, through all the modifications of colour and habit and performance, to the very end.

The actual arrival of a big run of salmon is a surprisingly subtle thing, in spite of the size of the fish and the mass of the schools. One can live right beside the mouth of a stream and hardly know that run has turned into it. I have sat in an anchored boat at the mouth of the Nimpkish River and watched school after school of cohos turn in through a whole afternoon and evening. The schools were big, sometimes hundreds of fish closely packed and only a few feet under the surface. Sometimes their passage showed in faint ripples on the surface, more rarely in swirls as they turned in some sudden panic. Sometimes a school turned downstream, turned again and came back. I

do not remember that a single fish jumped in the river channel, though there were occasional jumpers to the north, just before the schools turned in. For the most part I was watching only a narrow strip of water, perhaps a hundred feet wide, between the boat and the north edge of the channel; how many thousands of fish passed around me and behind me, I have no idea. But at the end of it all there was an odd sense of disbelief in the whole affair. There was little or no sign of fish on the rising ride upstream; there was only the visual memory of those hundreds of bright clean bodies, timid yet purposeful, slipping secretly into the river that had spawned them. Had there really been so many?

The Campbell is a broad, shallow stream where it passes my house. Thousands of salmon and steelhead run up through that stretch every year, yet I have scarcely ever seen one on its way. The large movements may well be at night and in daylight it would be natural enough for fish to pass stealthily through such an open, shallow reach, but it is still a surprise to find them suddenly in the pools farther upstream, a few hundred pink salmon towards the end of July, the big king salmon by the first week of August. Over the deep strong water of the pools they are not afraid to show themselves, the pinks breaking the surface like rising trout, the kings rolling out and shouldering over with a power that breaks the water white and starts echoes from the banks.

At this stage there is nothing confiding about the fish. They have their full strength; they are not ripe for spawning and they are keyed to protect themselves from whatever dangers may threaten. The silvery brightness of salt water is gone, but change towards spawning shape and color is gradual, almost imperceptible at first in these early fish. One fishes a fly over them and among them in search of steelhead, cutthroat or early-running coho, grateful for the magnificent way they show themselves, occasionally casting to cover one that is rolling persistently in the same place. Nothing comes of it except the rare, breathtaking coincidence of rolling side and drifting fly. But who is to say that nothing more than this is possible?

Later, with the cold, wet winds of early November, all this is changed. The fish are in the shallows now, active all across them. The gravel is loosened, freshly gray and brown where it has been turned by the tails of the females. Most of the pink salmon have spawned and died but kings and chums are everywhere, spawning, dying and dead. There are a few cohos among them and even one or two scarlet-bodied, green-headed creek sockeyes.

At such times I usually leave my rod on the bank and wade slowly upstream among the fish, pausing long and often to lean on my wading staff and watch. So long as I am slow and careful they do not rush away from me. It is easy to approach within reach of man's many tools of destruction, a little less so to approach within reach of the less deadly claws and teeth that are a bear's tools; from this threat they will rush a few yards upstream, throwing spray with their great broad tails. It is time then to stand still.

Any real concern for self-preservation has largely left them. They are obsessed with sexual purpose and the imminence of death leaves no leeway for other concerns. A month ago they would have started, arrow-swift, from a shadow. A month ago these shallows that they cling to with such urgency would have seemed places of terror to them. A month ago there was neither male nor female in their concerns. The new preoccupation is physical and mechanical, of course, but it is also ruthlessly logical even in its disregard for the dangers that may defeat it, because the time for fulfillment is so short. Successful spawning is the preservation of the race, within a month, spawned or unspawned, these strong bodies will be little more than a few scattered vertebrae and horny gill covers.

As I stand still in the hurrying water they settle back to their fierce pursuits and plungings, to gentle, questing swimming, to holding and swaying and shifting. Here and there a female shows her broad side in the fierce, flat thrust of nest digging. Great fish brush against my waders, even pass calmly between my legs. I am nothing to them unless an odd-shaped tree root caught on the shallows. Watching their eyes that neither see nor meet with mine, my mood tells me that they know me, know my concern for them and are not afraid. I do not have time then—nor want it—to remember that a bear waiting quietly to scoop them to death would find the same acceptance until he moved. A bear has his part there, inherited from his ancestors; but then I, too, am a rightful spectator, my way paid by understanding, my part to watch, to sympathize, to enjoy, to hope that there will never be less of them on these fall shallows than I count today.

Shallows like these are not the best place to watch spawning; the surface of the water is too much broken by its speed over the rocky gravels and light strikes from it in all directions, distorting shapes and colors, obscuring the detail of movement. Yet there is much to be seen. The king salmon females are usually rusty red, their males almost black. The bodies of both have lost much of their heavy-shouldered thickness; often they are scarred and blotched with destroying fungus. Male and female are by no means always easy to separate, except by their actions, a female coho may be as black of head, as crimson of body, with jaws almost as heavily hooked as her mate, but it is she that will dig the nest, not he. The pink salmon males are fantastically humped, most of them now dying, while the females hold a certain grace of shape in spite of their exhaustion.

Generally I move slowly upstream to a prow-shaped maple whose mossy trunk reaches horizontally ten or fifteen feet over the water before rearing up towards its crown. It is a good place to watch when the light is right. The water is too easy and shallow for the big kings, but chum salmon spawn directly under the tree and for fifty or a hundred feet upstream. Once I watched a small and lively female at work there, shadowed by five large males. The vivid black stripe of her side, edged with gold, showed clearly as she turned to stir the gravel. Then the males closed over her in a struggling mass and she forced from among them to break the surface and show that black and golden side in the air. The whole thing was repeated several times and I never did see the end of it or understand its real pattern and purpose. I have seen the shudder of spending fish close under the crooked maple, the lifting of their heads, the opening of their jaws, the clouding of milt in the bottom of the egg pocket. But in watching from above one can never see all the intricacies of the spawning act or understand fully the closely interdependent parts of male and female.

In the little streams one can see everything much more clearly, but the fish are less confiding—one must keep one's distance or they will start away and scatter. Even so, their colours are plain and the patterns of movement seem less haphazard and confusing. Occasionally, when two fish are paired apart from the others, it is possible to recognize the significant intricacies of behaviour that are fully revealed only in the observation tank.

Throughout the nest digging of the female, it is the male's part to stimulate her. He does this by continuously and closely circling over her so that his belly lightly touches her back. If his circling becomes too high, he expels one or two air bubbles from his swim bladder and sinks deeper; if he finds himself bearing too heavily upon her he rises to the surface, takes in more air and resumes his circling at the proper depth. From time to time, usually when she settles back into the gravel hollow to test its depth with her anal fin, he settles beside her, vibrating his body against hers, raising his head, opening his jaws, occasionally even shedding an involuntary jet of milt. Unless the female has decided the nest is ready, all this is without effect and she thrusts violently forward again on her side, throwing gravel back with her tail. When all is ready the two hold side to side, their vents closely over the depression in the gravel. Their bodies shudder powerfully, heads rise together, jaws open wide, curving bodies force deeply down into the egg pocket as eggs and milt are shed together. The whole process is repeated several times before the fish are finally spent and ready for death.

This death is no anticlimax, nor is it the inevitable consequence of spawning. Precocious male Pacific salmon that have never been to salt water, like their Atlantic counterparts, spawn successfully and do not die. It is not simply a matter of old age; Dr. O. H. Robertson of California has kept castrated sockeye salmon alive and healthy for seven and eight years, or about twice their normal life span. Yet it is a particularly complete and, in a sense, perfect death because everything about the fish—blood, tissues, organs, the whole body in all its parts—ages simultaneously. The salmon dies, not as man does, through the failure of some single part of him while the rest is healthy; he dies totally, his whole life force used up in the fulfillment of return and reproduction.

Dr. Robertson has shown that this is brought about chiefly by the extraordinary activity of the pituitary and adrenal glands that accompanies—and presumably in large measure controls—the maturing and spawning of a full-grown Pacific salmon. In other words the salmon's life is a gradual crescendo from the moment he turns towards his river and the violent activity of the spawning beds.

Even in the short-run rivers a few salmon die without spawning; if they are held away from the shallows by heavy floods, many may pass their time and die unspent. But under normal conditions there is a safety factor of several days between the covering of the last eggs in the final nest and death. Once I pitied the salmon in this state. Now I love them. They are death itself in a shell of life, but that remaining shell of life, though without hope or reason beyond the safety factor it provides,

is impressive. The fish conserve strength by sheltering in the bays and eddies of rivers and in the quieter water near the banks. Raccoons and bears find them easily there, the predatory sea gulls dive down at them until at last there is no longer strength to support them even in the lessened currents; they turn sideways, lose balance, drift and die.

But until that last strength is gone the will to live is still in them and finds its expression. At some too close threat of danger they will still drive away, upstream or down, in a surge of energy that seems no proper part of their wasted shapes. Bodies white with parasitic fungus, great king salmon somehow hold station over swift-running spawning shallows as though they still could play a part. I see them now on the shallows near my house, often two fish together, slowly forced down by the current, turning fiercely against it as it presses on their broad thin flanks and warns them of their weakness. It is the sort of thing man has glorified in himself as the undying spirit of man. Seeing it here so clearly, long after hope and purpose have gone, I can recognize it for what it is: the undying spirit of animals. I find it no less admirable.

Roderick Haig-Brown

The Caplin Are In

It began on a cool, misty morning when Pete, one of my nextdoor neighbours, came plodding up the road from Beachy Cove with a bucket in one hand, and paused at my gate to announce, "The caplin are in." He had perhaps a hundred of the little silver fish which his wife, Janet, would fry for their breakfast. "Just a few," he said. "They came on the spring tide last night."

"Any in now?" I inquired.

"No," he said, "but they'll be back with the next tide."

They were, too, several million of them, struggling in the inch-deep water at the very edge of the land. Each evening after that we waited on the beach, cast nets ready, for the coming of the tide. Then, as the sun set and the sea became milky and obscure, someone would exclaim excitedly, "Look there—right off the mouth of the brook!" And there they would be, packed together in unbelievable masses, turning and flipping and struggling in the gentle surf, then, as the surge receded, somersaulting breathless over the sand toward the retreating line of the sea.

We waded in to the harvest, men with cast nets, boys with dip nets, women and children with pans and tubs. Some of the younger boys even took off their clothes and waded out, shoulder-deep in the icy water, filling buckets or baskets with the small fish that swarmed about them in a squirming mass. My own net, made of hempen line with a circle of heavy lead weights, would take hundreds of the eight-inch fish—perhaps even a thousand—in a cast. The net, drawn ashore, would be too heavy to lift without danger of breaking, and we would have to ease it carefully up the beach before emptying it. In about ten minutes, that first big evening of the caplin run, everyone had all the fish he wanted, and two men were busy loading a dory with the surplus.

"'Tis a great mystery, isn't it," Peter commented. "They come like this and stay two or three weeks, and then they disappear, and there isn't a caplin to be found anywhere for a twelvemonth. No other fish is like that."

A mystery it is, but the mystery is inside the caplin, where it has been planted by the awesome forces of creative evolution. There is no mystery about where the caplin go or how they return. The mystery lies in the marvelous blueprint by which they operate—the complicated set of instruction which, to cover our ignorance, we call by the name of instinct.

"They say the caplin is a one-year fish," my friend Joe Harvey, a fisherman, remarked as we watched the unbelievable fury of the caplin scull at Beachy Cove. "Not many of them ever get back to sea, but the spawn is ready to make more spawn a year later."

All the fish we caught that first day were males. This is the usual thing, for shoals of male caplin regularly arrive first, and mill about in utter bewilderment. As many as a hundred million may be packed together along a quarter of a mile of beach without a single female in sight. This is the time when the fishermen go all out to get as many as possible, for the male fish are larger and have better flavour than the spawning females.

At a few beaches the females never show up at all. At others they arrive three or four days after the peak of the male run is past, and perhaps after more than half their potential mates have been transfered to trucks and carts and boats. They mate in threes, each female being conducted to the beach by two males, which are equipped with special little hooks on their fins for holding her between them. The females' eggs and the males' milt are deposited in clouds that turn the shallow water bluish white. As the tide retreats, the beach is spongy with spawn, and if you walk over it you can feel your feet sink, as on a carpet.

Some of the eggs hatch as they float on the water, well off shore, but most hatch on the wet part of the beach, the minute caplin being washed out by the next high tide to join the hosts of other larval creatures in the plankton layer at the surface of the sea, there to grow, if they are lucky, and join the breeding run another year, or, if they are not, to provide food for squid and herring, tiny dovekies and mighty whales, all of which skim and strain the top layer of the ocean for its rich, though invisible, hordes of minute animals and plants.

The fox family joined in the universal caplin feast that provided food for everything from sand fleas to men. By day, they would walk quietly in the woods within sight of the beaches as the level of the caplin scull rose to its afternoon climax. Men would come and go with cartloads of the small fish, and children would run whooping from one end of the beach to the other, following the surges of the spawning shoals.

But when the day's turmoil ebbed into darkness, they would step quietly out of the woods, tiptoe over the beach stones, and, at the edge of the water, in the white glare of sea and sand and moon, they would find the long piles of nickle-plated fish free for the taking. Then every fox had

a full belly. The savannah sparrows and the other ground-nesting birds were left in peace. For foxes, with their lifelong love of play, have much more interesting things to do than hunting food which they do not need.

Evening after evening, while I lay quietly on top of the bluff, scanning the shore with field glasses, they came in twos and threes, quiet as shadows, pausing to reconnoitre within the last line of shadows before venturing, a step at a time, to the exposure of the open beach. Then they would step out slowly, in single file, to begin their silent foraging among the stones and along the edge of the hard-packed sand just above the waterline.

The cubs were now so well-grown that they looked like rather lean and rangy adults, except when their mother was nearby for comparison. Then they seemed, suddenly, to be cubs again, shrunk to half their former size. Usually she preferred to forage alone, on her private section of the beach, keeping one wary eye on the welfare of her offspring, but not encouraging them to keep too close to her. The cubs were fast learning to be independent, to feed themselves without supervision, and to take their own precautions against danger.

They were still cubs just the same, and acted like cubs much of the time. After their stealthy approach to the beach, and their serious, almost furtive venture into the open, they would soon get used to their exposed position and begin to romp. If one cub found a choice morsel—perhaps a small but fat flounder among the stranded caplin— another would come scampering up to investigate. Sometimes there would be a short exchange of threats, each backing off with bared teeth. Sometimes they would try to snatch food slyly from each other, and there might be a sudden nip and a sparring match. Sometimes they would circle, warily, before breaking into a sudden lunge and a short chase over the tumbling stones.

An atmosphere of wariness, of overall caution, surrounded these nocturnal games, however. The young foxes seemed never to forget the human dooryards, perhaps with men or dogs, just a few hundred yards away, overlooking the cove where they played, and their squabbles were always conducted in near silence. Once or twice I caught a short, low growl, but even the foolhardy cubs never forgot themselves so far as to yelp or bark on the beach, as they did with complete freedom in the safety of the woods or on the mountain.

One evening as they romped in the moonlight, there was a sudden halt in their game, and they crouched low among the stones, backs flattened, brushes lowered, snouts pointed. Some signal of danger that I could not see had reached their keen ears or noses, and they were transformed instantly from carefree youngsters at play to cautious adults, ready to make full use of every nerve and muscle in defence of their lives.

Some minutes passed before I saw the big tom cat, streaked gray, and almost invisible against the gray stones. He had come out of the bushes at the foot of the bluff, and edged forward, silently, toward the stranded caplin along the high-tide mark. He continued to steal forward, infinitely cautious, ears flattened, his steps so slow you could barely see him move. He might have been stalking a bird or a mouse, but all he wanted, this time, was a share of the dead fish. He reached the tide line not far from the bluff, crouched there for a moment, facing the nearest fox, perhaps twenty yards away, then began gnawing a stranded caplin. The night was so silent that the crunching of his jaws could be heard from where I lay.

After a few more moments of suspicious watching, the young foxes accepted the cat, with reservations. Their games were ended for the evening, but they quietly resumed their feeding, never once turning their backs on the big tabby, or approaching him by so much as a step, but sharing the beach with him in a sort of armed truce.

So the days and the nights of the caplin scull brought their episodes of contrast and repetition, until, one morning, there was a strange silence along the beaches. Three weeks after the miracle of the caplin began, it was all over. Without the myriads of tumbling fish that we had come to expect, the surf, breaking on the foreshore looked empty. The sand and gravel were coloured with the golden spawn, but that, too, would vanish.

The eggs would take a lunar month to hatch—a fact suggesting that their hatching was timed, originally, by the movement of spring tides from moon to moon. Dropped at the peak of one tide, they would hatch with the coming of another.

Meanwhile, the surviving fish were moving back toward deep water. Those that had escaped the pursuing hordes of squid, cod, whales, those that had not been caught by kingfishers, puffins, or gulls, those that had not been taken on the beaches by foxes, cats, or above all, by humans, swam slowly, in beautiful formations, their small snouts in line, their tails flicking in unison, toward the open sea. There they sank slowly through the layers of sunlight toward the waters of midnight blue from which they came, to begin their annual renewal. In their own secret places of the ocean, where the currents bring masses of the minute plankton on which they feed, they would browse quietly and grow fat, and the spawn within them would ripen towards the next year's caplin scull.

The beaches that they left behind were littered with the debris of their living and of their dying. But the sea has its own ways of cleansing its shores. First came the gulls and terns in wheeling flocks, squabbling and calling, inviting their fellows to share the feast then trying to drive them off when they arrived. The tiny rock crabs and sand spiders ate the scraps. When the feast was done, a spring tide came in, washing the sand, the pebbles, the water-worn boulders, then retreating slowly past the line of the kelp on the offshore shallows. When dawn broke once more, the beach lay clean and empty and scoured, as though the drama of life and birth and death had never been played out on its rocks.

from *The Foxes of Beachy Cove*, by Harold Horwood

Waiter! . . . There's an Alligator in My Coffee

Waiter! . . . there's an alligator in my coffee.
Are you trying to be funny?
he said:
what do you want for a dime . . . ?
. . . a circus?
but sir! I said,
he's swimming
around
and around
in my coffee
and he might -
jump out on the table . . .
Feed him a lump of sugar! he snarled -
no! . . . make it two;
it'll weigh him down
and he'll drown.
I dropped two blocks of sugar
into the swamp
two grist jaws snapped them up
and the critter -
he never drowned.
Waiter! . . . there's an alligator in my coffee.
Kill him! Kill him!
he said:
BASH HIS BRAINS OUT
WITH YOUR SPOON . . . !
By this time
considerable attention had been drawn:
around my coffee
the waiters, the owner,
and customers gathered.
What seems to be the trouble?
the owner inquired,
and I replied:
There's an alligator in my coffee!
. . . But the coffee's fresh, he said
and raised the cup up to his nose . . .
Careful! I said,
he'll bite it
off
and he replied:
How absurd,
and lowered the cup
level to his mouth and
swallowed
the evidence.

Joe Rosenblatt

Anemones

Under the wharf at Saturna
the sea anemones
open their velvet bodies

chalk black
        and apricot
            and lemon-white

they grow as huge
and glimmering
        as flesh chandeliers

under the warped
and salt-stained wharf
    letting down
        their translucent mouths
                    of arms

even the black ones
have an aura
like an afterimage of light

Under our feet
    the gorgeous animals
        are feeding
                in the sky

Pat Lowther

Cataract

First everything turns to rainbows
edges of bronze and blue
doppler colour,
seen through a fine curtain
or the continual cast up
spray from a great fall.

Later the curtain thickens
white fog obscures shapes
hearing grows tense
for the rush and pressure
of blood like a great river
gathering volume
falling among caverns
in the listening skull

rushing toward the gorge
thunder and rub
to the precipice under the ear
to dive like Niagara
into the abyss
the hollow continent
the body

He is locked in
the white space
the mist   and the cataract
of blood he is forced
now to hear   and sways
like earth shaken
by its passage.

Pat Lowther

Hermit Crabs

In a pool maybe the size
of a man's forearm
there are hundreds of them,
little curled amber
snail shells scuttling sideways
like no snails.
You can just see
their brindle legs
fine as the teeth of a fine-tooth
tortoiseshell comb.
Five of them might
cover my fingernail,
but poke one
and he'll put out pincers
thin as bronze wire
and dare you
to do it again.

Pat Lowther

Boy Fishing at a Pier

Wisps of breeze
curl one way
then another
in air soaked yellow
with sunlight.

His hot skin
licks the wind;
thoughts, images
of the rotting harbour
slew slow
under a thatch
of hair hot to touch.

His legs dangle down,
small waves fritter at
the green-hung pile
crusted in spots
with barnacle villages.

His line wavers down
to the ripples, and
down into the drink-
ing swirl, fleck-
ering life, light.

Sharp thoughts of fishes.
A colour stabs one
and his guts thrill.

Sharp thoughts of fishes.
Leave the stupid conformist
his edifying dream
of others even stupider than himself

Sharp thoughts of fishes,
all life is light
earth and water,
but first light
then short darkness,

but spiralling up
atoms trickling
to light again
even thru saw-rip-
ping teeth staining
water briefly red,

to feed the sun
in the eye that
the sun in the sky
ignites to fill
the brain with light.

Taut, relaxed,
taut relaxed under
stretching gold
skin a muscle lives
as life chases, tears,
luring with colours, spins
smaller and smaller, dying
in the moment of delight.

A scaled body works,
the tail works,
the gills work
flooded with tastes of
the harbour drinking the sea,
nibbling the land,
tongue touching the cool
droplets of a splash.

A sweat-drop glistens,
rolls, catches
on a glittering hair.

Spine curved erect, his singing
skull's fluid balance, his
penis a wee rosebud,
its thrill withdrawn
softly, singingly
into his entrails,
light's joy travelling
miles of capillaries.

Milton Acorn

If you were a hundred millionth
of a centimetre tall and were trapped
inside an ice cube, this is what you'd see:
a six-sided hall of frozen water
molecules, stretching as far as
you could see.

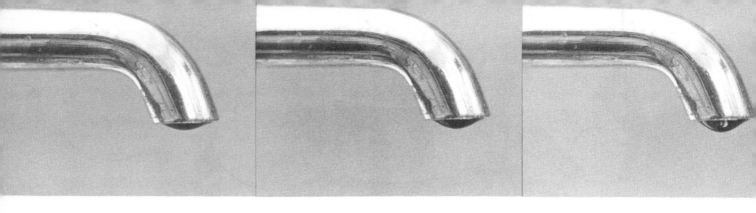

The Water Molecule

The water molecule is simple and uncomplicated as molecules go, but even a simple molecule is mysterious to us, because the molecular world is so different from ours. The water molecule you see here is magnified about one hundred million times—it's H_2O, two atoms of hydrogen bonded to the larger atom of oxygen. Straightforward enough—but "bonding" doesn't mean they're nailed or glued together. Instead they share some of the electrons that move around the nucleus of every atom. In water, it's an unequal sharing, as the electrons that used to belong exclusively to the pair of hydrogen atoms tend to spend more of their time around the oxygen, leaving the nucleus of the hydrogen atom exposed—a naked proton.

If you were a hundred million times smaller than you are, you could enter this molecular world and *feel* the effects of this unequal sharing. As you circle the water molecule, you'd feel the strong presence of positive electricity as you passed each exposed proton, but as you moved on, that would quickly fade, overwhelmed by the halo of negative electricity around the oxygen atom. The halo is really a swarm of negatively charged electrons, in-cluding the two pulled from the hydrogen atoms. Like you, other water molecules can "feel" these patches of positive and negative electricity on the surface, and where millions of water molecules are together, they arrange themselves to reduce the electrical tensions caused by these exposed charges.

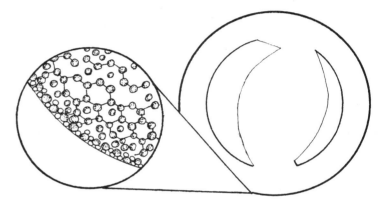

Ice Worms

Most ice cubes are streaked with fingers of air bubbles, called "ice worms". These long, thin bubbles tell you something about what's happening in the molecular world of that ice cube. Think back to when the water was poured into the ice cube tray. The water might have been cold, but it was still liquid. Billions and billions of water molecules were bumping and jostling and sliding past each other. But it became colder, and the molecules gradually lost their energy. They moved slower and slower, until they began to be trapped, locked onto the surface of the growing ice crystal. There's only one way any ice crystal can grow—that's by adding water molecules in groups of six, like building a giant honeycomb.

And it's so exact, so symmetrical, that no foreign atoms of any kind can be built in, because they'd disrupt the crystal that includes air. So as the ice crystal advances, air is squeezed out of it. Eventu-

ally there's enough air to form a bubble, but that crystals of ice on three sides, so there's only one way to go. Forward, away from the ice, and that's why you see long slender fingers of air in ice cubes.

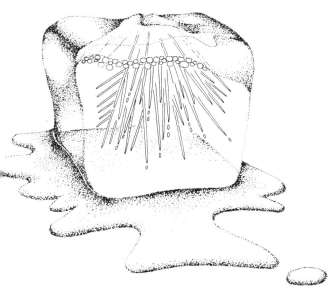

As the bubble grows, the ice follows, locking it in place forever. So look at the direction the ice worms point, and you can tell how the ice cube grew. The fingers were always pointing towards open water, with the ice following closely behind.

Snowflakes

Snowflakes, as we all know, have six sides. They have to. They're made of ice crystals, and ice crystals in turn are made up of water molecules frozen into place. And the only way water molecules can be locked together is in sixes: the corridor through the ice crystal a couple of pages back has six walls. All ice is built that way. Now if you can imagine a six-sided ice crystal, growing bigger and bigger in all six directions . . . and if you give it enough time, it will eventually be big enough that you can see it.

But even after all that growth, it'll still have its six-sidedness. That is a snowflake.

You might never see a photograph of the six-walled corridor through an ice crystal, but every snowflake you see tells you it's there. Sometimes snowflakes have long arms, like the one drawn here (this one was first drawn in 1872). It's easy for long arms to form, because as the snowflake drifts through the clouds, water molecules can hit the arms from any direction: above, below, or from the sides, so the arms can grow quickly. It's different if the snowflake is like a six-sided plate. Then only water molecules that hit the edge can be taken in—so a plate grows more slowly. But nobody really knows why there are so many different shapes of snowflakes.

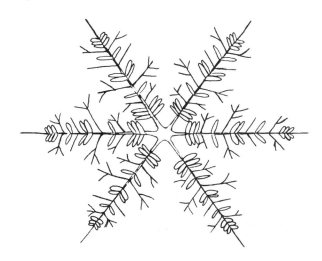

Is it true that there are no two snowflakes the same? The answer is. . . . possibly. You can't give a definite yes, because no one's going to look at all the snowflakes that fall. But with a few numbers you can show that it's possible that every snowflake that's ever fallen has been different: A typical snowflake weighs about a millionth of a gram. But if you measured all the snow that's fallen since the earth was created, it might be

85

something like sixty times the weight of the Earth! You'd think there must have been two identical flakes in all of that. But in each snowflake there are

1,000,000,000,000,000,000

molecules of water. The number of ways you could piece them together to make different snowflakes is unimaginable, but it's even a bigger number. So it is *possible* that every snowflake is different.

Glass of Water

Next time you have a drink of water, give it—the water—some thought. It isn't quite as uninteresting in there as you might believe. Liquid water is actually seething with activity. For one thing, it's not like ice with its row after row of neat, symmetrical six-sided spaces, like a three dimensional honeycomb. Liquid water's warmer, so the molecules are trembling and vibrating, so much that the perfect, six-sided crystals can't hold together. They begin to fall apart. Not completely—even as the

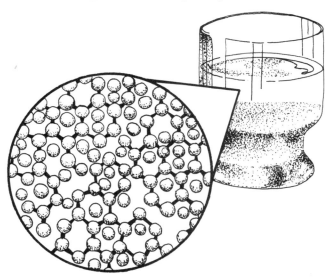

water slides down your throat, much of it is still in the honeycomb form. But in the glass, some of the crystals have broken down. These corridors that are completely empty in ice begin to fill up with stray molecules, jarred loose from other places. These strays slide and move down the corridor, pushed and pulled by the electrical forces from other molecules. Some of the crystals are sheared off, and fractures appear in the honeycomb, as large sections begin to slide past each other. But maybe the strangest thing is the movement of the hydrogen atoms. As the water molecules move and slide past each other, the hydrogens actually shift from one water molecule to the next—popping out of place from one, and into place in another. When one hydrogen atom does this, it forces another out of place. That in turn displaces a third, and suddenly the whole mass of water molecules is filled with a chain reaction, as this hydrogen shift moves like a zipper through the liquid.

Water Hole

If we Earthlings ever communicate with an extra-terrestrial civilization, it will probably be by microwaves. Radiation like this travels through space at the speed of light, and a message can be sent at the wavelength of your choice, just as we send messages on earth by microwaves and radio waves. The problem is, what wavelength to choose? We know where the radio stations are on the dial, but what wavelengths might a civilization at the other end of the galaxy be monitoring (listening to)? One guess might be a wavelength that any technological civilization should be familiar with. Something like the "song of hydrogen". Hydrogen atoms drifting in deep space emit radiation periodically as the single electron shifts energy levels in its course around the proton. The waves of this radiation are each twenty-one centimetres long, about eight and a half inches. Signals like this would arrive at 142 on your FM dial (if your FM dial went that far). Hydrogen is the most abundant

element in the universe, so everyone should know its song.

There's another wavelength that might be familiar to aliens: if you split a hydrogen atom from a water molecule, the fragment that's left is an oxygen atom bonded to a hydrogen atom. This pair emits radiation with waves eighteen centimetres long. The gap between H, alone, at twenty-one centimetres, and OH at eighteen centimetres is an ideal place in the microwave band to send interstellar messages. It avoids much of the noise radiating from our sky, the galaxy, and even the echoing radiation from the creation of the universe. Because this three centimetre space lies between the atoms that make up the water molecule, it's called the "water hole". As it is in Africa, it might be in the galaxy: different species gathering together at the water hole.

Surface Tension

If you think you were tense waiting for this drop of water to fall, what about the water? It's always tense! Water is tense at its surface because the molecules there are pulled tightly together as if the surface of the water had an elastic skin stretched over it. Insects can walk on the surface of water. You can even float a steel pin on water if you're careful. In fact, the surface of water is so tense that it alone supports the entire weight of a drop like the one that hung from the faucet above. Only when the weight of the water became too much did the elastic skin of the surface break and allow the drop to fall. Actually, if there were no surface tension, there would be no drops at all—the water would just gradually leak out of the faucet molecule by molecule and you'd never know it was leaking.

Why is the surface of water like this? You have to remember that water molecules attract each other quite strongly. *Inside* a drop of water, a molecule is surrounded on all sides by other molecules, so it's pulled from all directions at the same time.

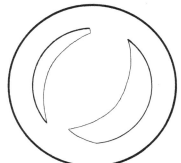

It tends to stay where it is. But at the surface of a drop, a molecule is being pulled by all the water molecules below it, but above it, there's only air. The downward pull isn't balanced by an upward pull, so all the surface molecules are pulled down, and pack together tightly at the surface. So tightly you can float a pin on them. In fact, if you look closely at a pin floating on water, you can see the surface of the water bending underneath the pin, like your mattress bends under you when you're sleeping.

Surface tension would have been a challenge for the Incredible Shrinking Man. While he could have walked on water, taking a shower would have been a different story. He would have had to tear and pull at the drops to break the surface tension so he could spread them over his skin. That is, if he survived the impact! Those drops would have been the same size as in our world, falling at the same speed.

Jay Ingram

87

Lee Side in a Gale

Black sea and shone-through sky
all mixed up along
a jumpity-jagged, beat-up
mercury saw of a skyline.

That old rusty cape hides me
but wind pokes round for me,
worries me like a scarecrow, howls
like a train from no-direction
then all-of-a sudden whacks me.

Milton Acorn

Offshore Breeze

The wind, heavy from the land, irons the surf
to a slosh on silver-damp sand.
The sea's grey and crocheted with ripples
but shadows, the backs of waves,
lengthen and lapse in the dim haze

The boats went out early, but now
come worm-slow through haze and distance;
the gunnels invisible, the men and engines
dots moving on a spit of foam.
They travel past my vision, past
that red jag of a headland, to harbour.

Milton Acorn

Dragging for Traps

When you're hanging to a pendulum
you wouldn't be there unless there's something to do
so mind the swing and mind your job;
like when you're out in a lobster boat
dragging for traps in the swell after the storm.
No time to think: "What am I
doing here, whose mother
loved me along with other fools?"

Turn into the waves and toss,
turn to the side and roll 45 degrees plus,
turn your back to them and mind the splash.
Just don't think you're going to be seasick.
All the time there's traps on the shore
bumped, bruised, broken, tangled with their lines.
Hold on and drag, trying not to be sure
that they're the very traps you're dragging for.

Milton Acorn

The Sea

Limitless sea? A lie! The sea's limited
To a variousness—nothing it's ever been can be repeated.
The light striking the water now, here, an hour before sunset
Makes it the colour of a pale Chinese ink
But such a precise shade of Chinese ink has never existed
Nor will any part of the sea be that precise shade again.

Infinity and finitude play loop-the-loop.
The sea, in every moment of all its ages
Has been different from anything it's been
And different from anything it'll ever be.
Repetition is impossible—no painting, no matter how
    truly
It catches the rage, the play, the calm of the sea
Is anything like the sea can possibly be
Yet there's no limit to its possibility
Even tho there're things the sea can never be.

No wave can ever duplicate another. The wavelets,
The weaving cross-hatches on the great waves
Can never cover one wave as they cover another;
The droplet of spray never flies from the same place
(how could you fix it as the same place?)
It rises. It roars. Like monstrous teeth
Its sudden upflingings threaten the atmosphere which
    torments it:
And its colours change, its loops and traps change.
    It'll never be the same.

Milton Acorn

People shouldn't get excited about reports of pollution of the oceans. Everybody knows that oil and water don't mix.

Phil Gagliardi,
BC cabinet minister

A Minute

I have to skate first. Impossible to finish the warm-up. Two minutes to go, and I stand by the boards trying to steel myself for the performance. Emptying my mind. I have to forget everything, everyone. Only the competition exists. . . . One minute to go, and I realize with horror that my skates are loose. They need to be tightened. There isn't enough time. My God! What shall I do? I bend over and tear at the laces. Frantic. They're frayed and won't pass through the eyelets. Panic thickens my fingers, renders them useless. I try to force the ends, clawing at them from the other side. My head reels. What can I do? Time is almost up. If I am late on the ice, once my name is called, I am disqualified. Yet to skate in loose boots could mean disaster. Desperation turns my fingers into pincers, and I somehow manage to lace my skates. But they still need tightening. They are still too loose. My name is called. I have no choice. I must go on.

Toller Cranston

Jackrabbit

March 12, 1975

Norway. Occupying the western and mountainous part of the Scandinavian peninsula, Norway is a land of natural splendour—pine-topped mountain ranges, valleys gouged out by glaciers, and narrow, deep-sided fiords. It is the land of the Vikings and it is inhabited by an intensely proud and self-reliant people.

Herman Smith-Johannsen was born in Norway on June 15, 1875. He travelled to many parts of the world, but finally settled in Canada, where, in the rugged terrain of the Laurentians, he found natural beauty to compare with his native land.

He is a small, wiry man, with intense blue eyes that hold you in a steady gaze as he speaks. This interview took place on a winter afternoon in early 1975 at his home. The questions were general so that he could be free to talk about those events in his life that he felt were significant.

His answers unfolded stories of life in a younger Canada, the history of skiing in North America, and, most delightful of all, a romance with the young daughter of a Cleveland judge.

You were born in Norway?
Yes, not far from the capital, Oslo. I spent most of my youth in the woods north of Oslo.

When did you start to ski?

I started to ski when I was two years old. They put me on a pair of skis. I tried to stand up and I fell down and I learned how to get up again.

Who taught you to ski?
You never took lessons to learn to walk. You don't need any lessons to learn to ski. It comes naturally. All you need to do is imitate someone who's doing it a little bit better than you. That's what I did. I watched some older boys.

Tell me about your family.
There were nine children. I was the oldest. My father's name was Fritz Anton Moritz Johannsen and he was a Kommandör-Kaptein in the Norwegian navy. Johannsen is a very common name in Norway and there were many Johannsens in the navy. So when he married my mother, whose name was Agnes Smith, he added her name to his to distinguish himself from the other Johannsens. I have one brother left now, and two sisters. My sisters are just young girls. One is 91 and the other

is 89. I remember my parents very well. They brought me up in the right way, although I haven't always been a good boy. Sometimes I took advantage of the life out of doors and I stayed away in the bush longer than I should have.

When you were growing up in Oslo, did a lot of people ski?

Oh yes, all my friends skied—cross-country and jumping and slalom. "Slalom" is a Norwegian word from the district of Telemark. It originally meant skiing around natural obstacles—big boulders and trees. At first, we didn't use any poles. It was all a natural way of living, a natural way of playing. When I was a very young fellow we often used only one pole on a long trip because we had rifles slung over our shoulders for hunting.

You went to school in Oslo?

Yes, I took great interest in my schooling. The teachers were wonderful. My geography teacher, for instance, was a fellow who had been all over the world. We had to draw maps of all the different countries in the world and he would tell stories that gave the places real life. The teacher who taught mathematics was also very much interested in skiing and, when a holiday was needed, he took us out to spend a night in the bush.

When did you first leave Oslo?

In 1894, when I graduated from the Military Academy in Oslo as a lieutenant in the regular army. I left the army in order to study engineering in Germany. The most important school in engineering and medicine was then in Germany. That was the University of Berlin.

How long did you stay in Germany?

I graduated as an engineer in 1899. By then I had connections with the European representatives of a company in Cleveland, Ohio, and I was introduced to the head man of the company in Cleveland. After I graduated, I came to Cleveland with the intention of eventually being sent back to Europe. But that didn't pan out.

Why did you want to go to the United States?

Well, no Norwegian leaves Norway because he doesn't like Norway. He leaves because he has a desire to see the rest of the world. That was my reason for going, always hoping that I'd be back in Norway sooner or later. But when you get married, have children and grandchildren, you become accustomed to other countries and finally you don't know where you belong. You're at home everywhere.

What did your job involve?

I was in the sales department, selling heavy machinery. I travelled a lot. They sent me to places nobody else wanted to go. One day they sent me to Canada. Nobody wanted to go to Canada. The Americans had funny ideas about Canada. They thought it was full of bears walking in the streets. They sent me up north to sell machinery to the Grand Trunk Railroad, later called the Canadian National. That company also sent me to Cuba and to Panama when the Canal was being built.

When did you first come to Canada?

About 1902. They sent me to North Bay. The railroad was then being built. Then farther north into Cobalt and Cochrane. I travelled through Quebec and later from Quebec to Winnipeg and out west across the Rockies. I had to go where they intended to run the railway to find out what kind of machinery would be used.

How did you travel into Canada in 1902?

Well, of course, there were no automobiles. Mostly by dog sled. I travelled by train to the end of the line, and then by horse and sleigh and finally by skis or on dog sled.

Was it at this time that you became involved with the Cree Indians?

Yes, the whole country north of North Bay was wilderness. When they found coal and silver at Cobalt, of course, the railroad pushed in farther. Previous to that time, there was nothing up there. When you got as far as Cobalt, all you could do was use horse and sleigh or dog sled. Or skis, of course. This was before the Transcontinental. The CPR was running through North Bay, Sudbury and up north of Lake Superior into Thunder Bay, Fort William and into Winnipeg. But north of there, there wasn't anything. The people who lived up in the bush at that time were all Indians. Later on, of course, the prospectors came in. But if you wanted to travel back in the bush, you had to travel with the Indians.

That was great country. It suited me because it was all natural. I wish I'd kept a diary at that time. Sometimes I was away a couple of weeks at a time.

Were you travelling on skis at this time?

Yes. When I first got up into that country, they were using snowshoes and I had skis. I would travel with the trappers over the traplines, you see, and I was on skis and they were on snowshoes. Of course, when you're on snowshoes, you can use both hands to handle the traps twenty miles long with only a shack here and there, then you're better with skis. I came back to the same places twenty years later, and they were running with skis and had snowshoes on their back. They would put the snowshoes on at the traps.

Do you still have friends among the Cree?

I still have friends among them. I've got many up at James Bay where the hydro plant is supposed to be built. I'm dead set against it. And that's not necessary! We ruined the Indians. Doing the wrong thing to them all the time. In the United States the theory was kill them off. We were better. In Canada we traded with them, gave them something to drink in order to make drunkards out of them. We should have helped them by getting pleasure out of their way of living and then gradually letting them take advantage of what we had to offer. Gradually. In Cleveland, Ohio, I was told that young men should go west, but they must beware of the Indians. There's only one good Indian, that's a dead Indian. And they were killed off during the Indian Wars.

But when I went on my first trip out west, I came back with an entirely different idea of the whole situation. I admired Chief Joseph of the Nez Percé in Idaho. I admired Sitting Bull of the Sioux. They were on the way to Canada for freedom. But they were massacred. And Custer, General Custer. He got what was coming to him. I had a great admiration at one time for the cavalry, because I'm a cavalry man myself. But I have no more sympathy for the cavalry because of what they did to the Indians.

It was in Cleveland that you met your wife.

That's right.

What was her name?

Alice Robinson.

How did you meet her?

Well, it was in one of the parks in Cleveland, probably Wade Park. Some young people were skating on a pond in the park. I was on skis in the hills nearby. When they saw this, they quit their skating and came over to see what these skis were all about. Nobody knew anything about skiing at that time. It was all absolutely new to them. They wondered what anybody could do on a pair of sticks with two poles. One of these people was my wife.

What kind of woman was she?

She was a very gentle lady. Her father was Thomas Robinson. He had been a doctor during the Civil War in the United States. After the war, he settled down to a law practice and finally he became a judge. His wife's name was Ella, and my wife was their only child. She was born in 1882. She had a ladies' schooling and she taught kindergarten for several years before I met her.

I came along and I was a foreigner and I had these weird skis on my feet. We married in 1907 and she came away with me to all the different places that I visited. She has been with me on horseback all through the island of Cuba. She came with me to logging and mining camps in all different parts of North America.

We spent our wedding night in Montreal. I was on a business trip and she came along. On that trip, I sold a locomotive to the CPR.

If, around 1906, these young skaters in Cleveland knew nothing about skiing, would it be accurate to say that you introduced skiing, in the form of cross-country, to North America?

Well, I was one of the fellows who introduced it. But the first man from Norway who had skis with him when he came to the United States was a man that they now call "Snowshoe Thompson". He came over on a sailing ship in 1837. I believe that he was on the sailing ship that was owned by my grandfather, Herman Smith. My grandfather also acted as captain of the ship. That's my

92

information, but I don't know whether that's absolutely correct. I think the name of the ship was the *George Washington*. About 1849, at the time of the gold rush in California, Snowshoe Thompson was carrying mail across the Sierras to the mining camps on skis.

How long did you remain with the Cleveland company?

In 1907, I quit and set myself up as a manufacturers' agent selling machinery in the same places where I used to sell for the Cleveland company. My wife and I spent most of our time between 1907 and 1911 in Cuba. Our first child was born in Cuba and we named her Alice Elisabeth. At that time I had business in Spitzbergen in northern Norway. So I took my wife and my daughter to Norway with me to spend some time with my family. You see, I wanted my daughter to get a good start in life.

How long did you remain in Norway?

I didn't stay in Norway. I had to come back and find my living in the different countries where I was operating. They remained there almost a year.

Where did you finally settle then?

When the First World War started, I decided to open an office in New York. I was still a manufacturers' agent and the business was mainly for export. I did very well during the war years. I sold a lot of machinery to Norway. But I never took a great liking to New York. It was too big. So at the end of the war, I turned over my business there to my youngest brother.

Otto?

Yes, Otto carried on with the business. That business is still running in the name of his son and grandson.

Did you leave New York when you turned over the business?

Yes, I took my family to Lake Placid. I wanted a place where they could get the outdoor life while I was travelling around attending to my business.

Your family must have been complete by then.

Yes, there were three children by this time. Alice was the first. My son Robert was next. Then my daughter Peggy. Peggy was named Ella Margaret, Ella for my wife's mother.

So the five of you went to Lake Placid?

Lake Placid was a very convenient place for my family to grow up. At that time, at the Lake Placid Club, that was really the start of skiing in New York, in the eastern United States. After the first

war, my principal work was in Canada. I always had my love for the Laurentian Mountains. That's been with me ever since I first came to Canada. So I decided to set up an office in Montreal.

When was that?

That was after the First World War. 1919.

Did your family move to Montreal with you at that time?

No, I kept the family down at Lake Placid until 1928, when I moved them to Montreal. I drove to Montreal every Monday morning. Well, not every Monday, because sometimes I went way back into the bush. At that time I had a 1916 Winton Six. It was a big sedan. You could pile everything into it—skis, kids, dogs. Every Friday evening I would drive back to Lake Placid. That was 120 miles. My family and I would go camping, or skiing over the weekend. Then, on Monday morning, I would head back for Canada.

Tell me what Montreal was like at that time.

It was a wonderful town—a small town. All the English-speaking people knew one another. Nobody lived on the other side of the mountain, the north side of the mountain. We skied on Mount Royal, coming down Peel Avenue, past the Windsor Hotel, down to the Queens Hotel at Bonaventure Station. That was a regular ski run, in the old days.

When you first came to Montreal, then, people were skiing?

Yes, I still have a badge from 1905 from the Montreal Ski Club. That club was started in 1903. There were just a few people skiing. Skiing was brought to Canada by Norwegians. When I first came to Montreal, there was a consul here from Norway, and I think he and I were the only Norwegians in town. Most of the others at that time went out west.

You laid out the Maple Leaf Trail, didn't you?

Yes. I worked on that in the early 1930s. The Maple Leaf Trail started at Labelle and ran for 80 miles down to Shawbridge.

How long did it take you?

A couple of years. I had to go to all the farmers and make friends with them, stay overnight with them. This was after the Depression. They gave me a pass on the railway. They knew what I was doing to develop the country. Then I laid the Trail from the different hotels. So I could always get something to eat at the hotels. And I went hunting. I had moose and deer hanging in the wood shed so the family always had something.

Were you living in Montreal while you laid out the trail?

No, at Shawbridge, just about three miles down from Piedmont.

When did you move from Montreal to Shawbridge?

At the Depression. When the Depression came in 1929, I was wiped out. And since then I have never really gone back to business. I've had almost forty-five years of real life. But I owe everything to my wife and three children. If I hadn't had my wife and my three children cheering me up at that time, making it possible for me to lead a simple life out in the bush away from the town, I never would have been able to make it. When I think of the way we started. It was a very small shack, in Shaw-bridge. It was built for summer. And I had to bank it up with dirt and snow. And no automobile, no telephone, and ice instead of a refrigerator. Just as simple a life as possibly could be. It brought us closer together. Those days, I could go hunting. I always had a deer or a moose hanging in the woodshed and we lived on next to nothing. I had the knowledge of skiing which I used as a means of having something to give to other people. If you have something that you can do for other people, there's nothing like it. Then it all comes back to you. The reason I'm getting along now, and why they have me go here and there, is because I have done something for skiing and helped people take advantage of the sport without making money on it.

We lived a natural life and the family was brought closer together due to the hardship. So don't worry about hardships. It has a great influence on your life and you're profiting by it. I'm thankful for the good fortune that I had by being knocked out. I got away from the towns and back into real life, the same kind of life that I had when I was a boy in Norway.

You're very fortunate. Many people went under during the Depression.

A great many of my friends committed suicide. Jumped out of windows.

What made you decide to stay in Quebec?

Well, I had some of the most interesting times in my life when I was sent from Cleveland to northern Ontario and Quebec. I took a liking to Quebec. It's more cosmopolitan. Ontario is too American. There's freedom in Quebec on account of the two languages. Of course, the United States has only one language. In the United States an immigrant has to forget the past and become American. The company in Cleveland sent me to Canada by mistake, you see. And they never really got me out of Canada. And, of course, I loved the Laurentians.

What kinds of changes have you seen in Canada since you first arrived here?

I have seen Montreal grow from a cozy old-fashioned place where you got a lot out of living into an industrial town of skyscrapers. But Canada to me is north of Tremblant. Canada to me is the wilderness and there's no country in the world that has that to offer, the equal of Canada. It's the way of wilderness life, fishing, hunting, travelling with a canoe in the summer and with cross-country skis over the same portages in winter. You can't beat Canada on that.

But people who live in Canada today don't realize the wonderful country they have. You have to go back north into the wilderness to really discover it and make use of that way of living, particularly in January, February, and March, the best time of the year. You can build a fire way back in the bush somewhere and sleep in your sleeping bag in the snow. You don't even need a tent when it's cold enough. When it gets warmer then you need a tent or you get wet. But this time of the year, most of the time the air is dry up north and you're cozy with a big fire. You take a snow bath, and you feel like a million dollars. That's the life.

What do you think will happen to this way of life in the future?

We have plenty of territory. Canada is the only place in the world where you still have this kind of wilderness. But we've got to take care of that wilderness. We must get our politicians away from the idea of making money out of the wilderness and gradually develop it in such a way that we still keep the wilderness. It can be done. Our forest industry, our mines, and all the rest of the development, including the water-power development, they've got to be kept under control or they'll ruin the whole country and it won't be good for anybody. No matter how big the country is, it can be ruined.

Do you think it will be ruined?

I don't think so. I think that people have gradually come to the conclusion that we have to take care of our resources so that we leave the country in good shape for the next generation.

What do you think of the generation that will inherit this country?

The young people are all right. I admire the young people because they have ideas. But they must have something to offer to take the place of what we already have. These young fellows haven't got that yet. But the middle-aged fellows, what I call the status quo fellows, the people who want to keep what we have, they're wrong too because our social life and regulations can be improved upon. Some of the young people have ideas about that, but they don't know what to do with their ideas. The middle-aged fellows should be willing to discuss these things with the younger people to find out what can be done by using the new ideas and keeping some of what we have already developed. The old fellows like myself, now, I can say anything I want because I have nothing to lose.

I understand that you were installed as a Member of the Order of Canada.
Yes. I was made a Member of the Order of Canada in December, 1972, by Roland Michener. The last time I skied with Roland Michener was some years ago. It was the last stretch of the Marathon before you get into Ottawa, and I fell and broke my back. He beat me. When a governor general beats you, it's time to quit skiing.

I'm an honorary member of a whole lot of ski clubs. King Olav gave me a medal in 1972 and Sir George Williams University gave me an honorary doctor of laws. That was in 1968. I got an award down in New York on February 13, 1975—the Dubonnet Skier of the Year. So, you see, I've been pretty busy, but I don't deserve any of it. The only thing I've done is try to get young people into the right way of living by using the sport of skiing in such a way that they, all people, no matter how poor, get health and happiness out of it without spending too much money.

Do you still ski every day?
I have a two-mile run back of the house here and I'm over that practically every day, and I ski to the Post Office each day to get my mail.

In thinking about the many things you've done and the many awards and honours that have come your way, what would you consider the most thrilling moment in your life?
Well, I married the right girl. Without my wife, I wouldn't have been what I am today. I owe everything to her.

from *Jackrabbit: His First Hundred Years*

Peking Captures World Hockey Title

STOCKHOLM — Before the largest audience ever to watch a sporting event, Peking edged the NHL champion Vancouver Canucks last night to give China its first world hockey championship.

The Number One District People's Team, facing elimination for the third game in a row, handed the Canadians a 3-1 loss before a crowd of 55,950 and an estimated 450 million television viewers.

"A lot of work went into this tournament," said Hockey Canada president Alan Eagleson. "I think we've proven that hockey is an international sport—this game was broadcast live in eighteen countries. Commercially it was a success.

"But it burns me up to see a series this important ruined by bad officiating. The NHL will have to re-think its participation next year if the system of choosing officials isn't looked at pretty carefully. The Canucks should have won it all. They got a raw deal out there, let's face it."

"I'm proud of the team," said Canucks owner Bobby Orr. "A lot of people thought the North Stars should have been here, but I think we proved that beating them was no fluke. We ran into some bad calls, otherwise who knows? It seems like every time we play outside North America we run into this."

The game was controversial from start to finish. With the first period barely underway, Lin Fun's clearing pass deflected off the referee's skate into the centre ice area. No one was more surprised than Kevin McCarthy, who found himself with a clear path to the goal. He was hauled down from behind by Fun, but play carried on.

Vancouver coach Bobby Clarke protested so vehemently that referee Vladimir Popov assessed the Canucks a bench minor. The game was delayed for several minutes as Clarke threatened to pull his team off the ice.

Shortly after play resumed, Rick Blight had a golden opportunity to give Canada the lead. The veteran right winger picked up his own rebound and had an empty net to shoot at. As he has done so often in the series, C.K. Yang came up with a brilliant save.

The Canucks scored late in the period, but the goal was called back. After bobbling Mike Gartner's slapshot, Yang dropped to his knees to cover up. The puck squirted loose and Bruce Boudreau, parked on the doorstep, poked it home. Though the red light went on, Popov ruled that the

whistle had blown.

The Canucks took a 1-0 lead on Terry Lindsay's solo effort at 2:18 of the second period. Peking came back to tie it up less than a minute later, when Wing Wong took advantage of a miscue in front of the net and beat Pat Riggin for his fifth goal of the series.

Riggin was again outstanding in goal, stopping sixteen shots in the second period, including key saves on Peng Chan and You Lee. At the other end of the ice, the Canucks' only near-miss came late in the period with Peking a man short. Vancouver captain Denis Potvin's drive from the point hit the post. Mark Napier was unable to control the rebound with Yang out of position.

The Chinese came out flying in the third period, forcing the Canucks to take three penalties in a row. Only the heroics of Riggin kept them in the game. At one point he stopped four successive shots from pointblank range.

The Canucks, playing their eighteenth game in twenty-four days, had no answer for the persistent forechecking and precision passing of the Chinese. Wong's line in particular dominated play whenever it was on the ice. Vancouver was called for icing six times before the teams changed ends midway through the period.

At the twelve-minute mark, Potvin was carried off on a stretcher after tumbling head-first into the goal post. "It looked like he sort of lost his balance," said Riggin. "Lum was cutting in on the net, but Denis got back. I didn't really see what happened because I was watching the play. He was going full speed, though, and I could tell he hit the post pretty hard. He went completely limp. At first his eyes were open, but he didn't know where he was." Potvin was taken to hospital with undetermined injuries.

Potvin's misfortune seemed to spark the Canucks. When Chiao Ng went off for hooking at 14:06, the Vancouver power play functioned as well as it has all series. But it was Yang's turn to perform miracles. He robbed Mike Gartner from in close and made a fine skate save on Napier's screened drive from the point.

Then the power play backfired. Wong intercepted Napier's centring pass just before Ng stepped out of the penalty box, and the Canucks were caught with one man back. Wong made no mistake, drawing Kevin McCarthy out of position before sliding the pass to Ng.

Riggin moved well out of the net expecting a shot, but Ng elected to deke and had little trouble tucking the puck behind the Vancouver netminder. It was the first time Peking had been ahead in the series.

The Canadian players argued to no avail that Ng had been offside at the blueline. An incensed Bobby Clarke again threatened to withdraw his team, and only after World Hockey League officials negotiated with Clarke and Hockey Canada president Alan Eagleson did the game resume.

Vancouver went all out in an effort to get the equalizer, but Wang, named the tournament's most valuable player, was equal to the occasion. With a minute left in regulation time, Riggin was pulled in favour of an extra attacker. The Canucks did everything but send the game into overtime. Then, at 19:51, Wong's rink-length shot found the empty net, sealing the game and China's first world hockey title.

Peking coach Tai Dop Cheung called his team's victory gratifying and well-earned. "We believed we would win," Cheung said through an interpreter. "At no time during the tournament did we lose faith. We were behind in the series against Moscow also. Hockey is a collective sport. Each player contributed equally to the victory."

Cheung was asked if he had made strategy changes after losing three of the first four games. "The Canadian team is strong," he said. "We knew to win we must prevent goals. Opportunities arise when a team plays according to plan. Our players knew they must concentrate on defence. The Canadians have many outstanding scorers. To win we must stop them. This was the strategy at all times."

NHL president Ken Dryden expressed disappointment, but was quick to credit the Chinese. "There's no question that they've improved considerably," said Dryden. "We saw that at Prague last year. We saw it here in their victories over Moscow and West Berlin. The Peking team skates as well as any team I've seen.

"If they have a weakness, it's the work of their defencemen in their own end. But what they lack in size, they seem to make up for in mobility. Offensively, they have no weaknesses that I can see. And Yang played superbly tonight, as he has all series. I'd say that he made the difference.

"I'm personally disappointed, of course, both at the outcome and at the number of disputable calls. But we scored only twelve goals in seven games, and it simply wasn't enough."

Clarke, who has been critical of the officiating throughout the series, was less charitable in his assessment. "The guys play their hearts out, and we lose it like this. Nobody expected us to get this far, but the guys came up with a big effort. When we came back to beat Oslo, that gave us a real boost. But how can you win against a team like Peking *and* the referee? I've seen brutal refereeing, but this was a joke.

"Losing's never easy. But losing this way is really tough. I'm proud of the team. The guys have got nothing to be ashamed of. I'm not saying Peking didn't play a great game, but in twenty years of hockey I've never seen so many bad calls. No way."

"Sure I'm upset," said defensive coach Tom Bladon. "We have them on the ropes and they come back to beat us. Maybe we got a little too confident after winning the first two.

"We had a good first period, but Yang kept them in there. In the second and third we came out flat. I guess they were just hungrier than we were. There were some bad calls, but we missed a lot of scoring opportunities. You've got to take advantage of your opportunities. I don't mean to take anything away from Yang, but we were our own worst enemy.

"We let up, and you can't let up against a team like Peking. I thought we outplayed them overall. I still think we're a better team. But that's not what the record book will say."

Weekend, 1977

Heart Like a Wheel

Some say a heart is just like a wheel
When you bend it you can't mend it
And my love for you is like a sinking ship
And my heart is on that ship out in mid-ocean

They say that death is a tragedy
It comes once and it's over
But my only wish is for that deep dark abyss
'Cause what's the use of livin' with no true lover?

When harm is done no love can be won
I know it happens frequently
What I can't understand—O please God hold my hand—
Why it should have happened to me

And it's only love
and it's only love
That can wreck a human being
and turn him inside out
That can wreck a human being
and turn him inside out

Some say a heart is just like a wheel
When you bend it you can't mend it
And my love for you is like a sinking ship
And my heart is on that ship out in mid-ocean

And it's only love
and it's only love
and it's only love
and it's only love

Anna McGarrigle

from The Great Bear Lake Meditations

A man, warmly dressed, in perfect health, mushing
his dogs a short distance between two villages,
never arrives. He has forgotten to reach down,
catch a little snow in his mitten and allow it to melt
in his mouth. For a reason neither he nor his dogs
understand, he steps from the runners of his sled,
wanders dreamily—perhaps warmly,
pleasantly—through the wide winter, then sits to
contemplate his vision, then sleeps. The dogs tow
an empty sled on to the place at one of the two villages where they're usually fed. While those who
find the frozen man suspect the circumstances of
his death, always they marvel that one so close to
bed, warmth, food, perhaps family, could stray so
easily into danger.

J. Michael Yates

Thaw (for Donna)

Hearing you softly crying
and sensing your slightly
sullen disappointment
over the long distance line
I ventured out in sub-zero
January and begged
a truck driver
who didn't trust my looks
enough to let me ride in the cab
to at least carry me
in the open box
and was gratified to hear
the driver say to his helper
(28 miles from Saint John
shouting over a country & western radio)
"Well, he's tougher than he looks,
he's still with us."
But the cold, the cold,
I couldn't have been in the truck,
I was somewhere out front
suspended in a silver tunnel
the headlights were thrusting
through the monotonous darkness
of the New Brunswick night;
I might as well have been
a minute Jonah in a salmon's belly
being carried under the ice
all the way to Fredericton.

Oh, how I cursed you for your loneliness
and called myself stupid
for falling toward your distant tears,
I cursed the black woods away,
cursed away the powder snow
wisping past the tail gate,
cursed away the miles . . .
"All right mac, time to get out."
How did I walk the last five blocks?
My feet felt like my legs
were cut off at the knees,
my hands (being curled fists
inside my dandified dress gloves)
couldn't open the back door
so that I elbowed the window
or perhaps even banged it with my head
till you awoke and came
looking so warmly sleepy
and mildly awed and amazed
that I laughed at your silliness
as we got into bed
and I thawed inside your embrace
and again laughed at myself
for loving you so much.

Terry Crawford

By the River

But listen, she thinks, it's nearly time.

And flutters, leaf-like, at the thought. The train will rumble down the valley, stop at the little shack to discharge Styan, and move on. This will happen in half an hour and she has a mile still to walk.

Crystal Styan walking through the woods, through bush, is not pretty. She knows that she is not even a little pretty, though her face is small enough, and pale, and her eyes are not too narrow. She wears a yellow wool sweater and a long cotton skirt and boots. Her hair, tied back so the branches will not catch in it, hangs straight and almost colourless down her back. Some day, she expects, there will be a baby to play with her hair and hide in it like someone behind a waterfall.

She has left the log cabin, which sits on the edge of the river in a stand of birch, and now she follows the river bank upstream. A mile ahead, far around the bend out of sight, the railroad tracks pass along the rim of their land and a small station is built there just for them, for her and Jim Styan. It is their only way in to town, which is ten miles away and not much of a town anyway when you get there. A few stores, a tilted old hotel, a movie theatre.

Likely, Styan would have been to a movie last night. He would have stayed the night in the hotel, but first (after he had seen the lawyer and bought the few things she'd asked him for) he would pay his money and sit in the back row of the theatre and laugh loudly all the way through the movie. He always laughs at everything, even if it isn't funny, because those figures on the screen make him think of people he has known, and the thought of them exposed like this for just anyone to see embarrasses him a little and makes him want to create a lot of noise so people will know he isn't a bit like that himself.

She smiles. The first time they went to a movie together she slouched as far down in the seat as she could so no one could see she was there or had anything to do with Jim Styan.

The river flows past her almost silently. It has moved only a hundred miles from its source and has another thousand miles to go before it reaches the ocean, but already it is wide enough and fast. Right here she has more than once seen a moose wade out and then swim across to the other side and disappear into the cedar swamps. She knows something, has heard somewhere that farther downstream, miles and miles behind her, an Indian band once thought this river a hungry monster that liked to gobble up their people. They say that Coyote their god-hero dived in and subdued the monster and made it promise never to swallow people again. She once thought she'd like to study that kind of thing at a university or somewhere, if Jim Styan hadn't told her Grade 10 was good enough for anyone and a life on the road was more exciting.

What road? she wonders. There isn't a road within ten miles. They sold the rickety old blue pickup the same day they moved onto this place. The railroad was going to be all they'd need. There wasn't any place they cared to go that the train, even this old-fashioned milk-run outfit, couldn't take them easily and cheaply enough.

But listen, she thinks, it's nearly time.

The trail she is following swings inland to climb a small bluff and for a while she is engulfed by trees. Cedar and fir are dark and thick and damp. The green new growth on the scrub bushes has nearly filled in the narrow trail. She holds her skirt up a little so it won't be caught or ripped, then runs and nearly slides down the hill again to the river's bank. She can see in every direction for miles and there isn't a thing in sight which has anything to do with man.

"Who needs them?" Styan said, long ago.

It was with that kind of question—questions that implied an answer so obvious only a fool would think to doubt—that he talked her first out of the classroom and then right off the island of her birth and finally up here into the mountains with the river and the moose and the railroad. It was as if he had transported her in his falling-apart pickup not only across the province about as far as it was possible to go, but also backwards in time, perhaps as far as her grandmother's youth or even farther. She washes their coarse clothing in the river and depends on the whims of the seasons for her food.

"Look!" he shouted when they stood first in the clearing above the cabin. "It's as if we're the very first ones. You and me."

They swam in the cold river that day and even then she thought of Coyote and the monster, but he took her inside the cabin and they made love on the fir-bough bed that was to be theirs for the next five years. "We don't need any of them," he sang. He flopped over on his back and shouted up into the rafters. "We'll farm it! We'll make it go. We'll make our own world!" Naked, he was as thin and pale as a celery stalk.

When they moved in he let his moustache grow long and droopy like someone in an old, brown photograph. He wore overalls which were far too big for him and started walking around as if there were a movie camera somewhere in the trees and he was being paid to act like a hillbilly instead of the city-bred boy he really was. He stuck a limp felt hat on the top of his head like someone's uncle Hiram and bought chickens.

"It's a start," he said.

"Six chickens?" She counted again to be sure. "We don't even have a shed for them."

He stood with his feet wide apart and looked at her as if she were stupid. "They'll lay their eggs in the grass."

"That should be fun," she said. "A hundred and sixty acres is a good-size pen."

"It's a start. Next spring we'll buy a cow. Who needs more?"

Yes who? They survived their first winter here, though the chickens weren't so lucky. The hens got lice and started pecking at each other. By the time Styan got around to riding in to town for something to kill the lice a few had pecked right through the skin and exposed the innards. When he came back from town they had all frozen to death in the yard.

At home, back on her father's farm in the blue mountains of the island, nothing had ever frozen to death. Her father had cared for things. She had never seen anything go so wrong there, or anyone have to suffer.

She walks carefully now, for the trail is on the very edge of the river bank and is spongy and broken away in places. The water, clear and shallow here, back-eddies into little bays where cattail and bracken grow and where water-skeeters walk on their own reflection. A beer bottle glitters where someone, perhaps a guide on the river, has thrown it—wedged between stones as if it has been there as long as they have. She keeps her face turned to the river, away from the acres and acres of forest which are theirs.

Listen, it's nearly time, she thinks. And knows that soon, from far up the river valley, she will be able to hear the throbbing of the train coming near.

She imagines his face at the window. He is the only passenger in the coach and sits backwards, watching the land slip by, grinning in expectation or memory or both. He tells a joke to old Bill Cobb the conductor but even in his laughter does not turn his eyes from outside the train. One spot on his forehead is white where it presses against the glass. His fingers run over and over the long drooping ends of his moustache. He is wearing his hat.

Hurry, hurry, she thinks. To the train, to her feet, to him.

She wants to tell him about the skunk she spotted yesterday. She wants to tell him about the stove, which smokes too much and needs some kind of clean-out. She wants to tell him about her dream; how she dreamed he was trying to go into the river and how she pulled and hauled on his feet but he wouldn't come out. He will laugh and laugh at her when she tells him, and his laughter will make it all right and not so frightening, so that maybe she will be able to laugh at it too.

She has rounded the curve in the river and glances back, way back at the cabin. It is dark and solid, not far from the bank. Behind the poplars the cleared fields are yellowing with the coming of fall but now in all that place there isn't a thing alive, unless she wants to count trees and insects. No people. No animals. It is scarcely different from her very first look at it. In five years their dream of livestock has been shelved again and again.

Once there was a cow. A sway-backed old Jersey.

"This time I've done it right," he said. "Just look at this prize."

And stepped down off the train to show off his cow, a wide-eyed beauty that looked at her through a window of the passenger coach.

"Maybe so, but you'll need a miracle, too, to get that thing down out of there."

A minor detail to him, who scooped her up and swung her around and kissed her hard, all in front of the old conductor and the engineer who didn't even bother to turn away. "Farmers at last!" he shouted. "You can't have a farm without a cow. You can't have a baby without a cow."

She put her head inside the coach, looked square into the big brown eyes, glanced at the sawed off horns. "Found you somewhere, I guess," she said to the cow. "Turned out of someone's herd for being too old or senile or dried up."

"An auction sale," he said, and slapped one hand on the window glass. "I was the only one there who was desperate. But I punched her bag and pulled her tits, she'll do. There may even be a calf or two left in her sway-backed old soul."

"Come on, bossy," she said. "This is no place for you."

But the cow had other ideas. It backed into a corner of the coach and shook its lowered head. Its eyes, steady and dull, never left Crystal Styan.

"You're home," Styan said. "Sorry there's no crowd here or a band playing music, but step down anyway and let's get started."

"She's not impressed," she said. "She don't see any barn waiting out there either, not to mention hay or feed of any kind. She's smart enough to know a train coach is at least a roof over her head."

The four of them climbed over the seats to get behind her and pushed her all the way down the aisle. Then, when they had shoved her down the steps, she fell on her knees on the gravel and let out a long unhappy bellow. She looked around, bellowed again, then stood up and high-tailed it down the tracks. Before Styan even thought to go after her she swung right and headed into bush.

Styan disappeared into the bush, too, hollering, and after a while the train moved on to keep its schedule. She went back down the trail and waited in the cabin until nearly dark. When she went outside again she found him on the river bank, his feet in the water, his head resting against a birch trunk.

"What the hell," he said, and shook his head and didn't look at her.

"Maybe she'll come back," she said.

"A bear'll get her before then, or a cougar. There's no hope of that."

She put a hand on his shoulder but he shook it off. He'd dragged her from place to place right up this river from its mouth, looking and looking for his dream, never satisfied until he saw this piece of land. For that dream and for him she had suffered.

She smiles, though, at the memory. Because even then he was able to bounce back, resume the dream, start building new plans. She smiles, too, because she knows there will be a surprise today; there has always been a surprise. When it wasn't a cow it was a bouquet of flowers or something else. She goes through a long list in her mind of what it may be, but knows it will be none of them. Not once in her life has anything been exactly the way she imagined it. Just so much as foreseeing something was a guarantee it wouldn't happen, at least not in the exact same way.

"Hey you, Styan!" she suddenly calls out. "Hey you, Jim Styan. Where are you?" And laughs, because the noise she makes can't possibly make any difference to the world, except for a few wild animals that might be alarmed.

She laughs again, and slaps one hand against her thigh, and shakes her head. Just give her—how many minutes now?—and she won't be alone. These woods will shudder with his laughter, his shouting, his joy. That train, that kinky little train will drop her husband off and then pass on like a stay-stitch thread pulled from a seam.

"Hey you, Styan! What you brought this time? A gold brooch? An old nanny goat?"

The river runs past silently and she imagines that it is only shoulders she is seeing, that monster heads have ducked down to glide by but are

watching her from eyes grey as stone. She wants to scream out "Hide, you crummy cheat, my Coyote's coming home!" but is afraid to tempt even something that she does not believe in. And anyway she senses—far off—the beat of the little train coming down the valley from the town.

And when it comes into sight she is there, on the platform in front of the little sagging shed, watching. She stands tilted far out over the tracks to see, but never dares—even when it is so far away—to step down onto the ties for a better look.

The boards beneath her feet are rotting and broken. Long stems of grass have grown up through the cracks and brush against her legs. A squirrel runs down the slope of the shed's roof and yatters at her until she turns and lifts her hand to frighten it into silence.

She talks to herself, sings almost to the engine's beat—"Here he comes, here he comes"—and has her smile already as wide as it can be. She smiles into the side of the locomotive sliding past and the freight car sliding past and keeps on smiling even after the coach has stopped in front of her and it is obvious that Jim Styan is not on board.

Unless of course he is hiding under one of the seats, ready to leap up, one more surprise.

But old Bill Cobb the conductor backs down the steps, dragging a gunny sack out after him. "H'lo there, Crystal," he says. "He ain't aboard today either, I'm afraid." He works the gunny sack out onto the middle of the platform. "Herbie Stark sent this, it's potatoes mostly, and cabbages he was going to throw out of his store."

She takes the tiniest peek inside the sack and yes, there are potatoes there and some cabbages with soft brown leaves.

The engineer steps down out of his locomotive and comes along the side of the train rolling a cigarette. "Nice day again," he says with barely a glance at the sky. "You makin' out all right?"

"Hold it," the conductor says, as if he expects the train to move off by itself. "There's more." He climbs back into the passenger car and drags out a cardboard box heaped with groceries. "The church ladies said to drop this off," he says. "They told me make sure you get every piece of it, but I don't know how you'll ever get it down to the house through all that bush."

"She'll manage," the engineer says. He holds a lighted match under the ragged end of his cigarette until the loose tobacco blazes up. "She's been doing it—how long now?—must be six months."

The conductor pushes the cardboard box over against the sack of potatoes and stands back to wipe the sweat off his face. He glances at the engineer and they both smile a little and turn away. "Well," the engineer says, and heads back down the tracks and up into his locomotive.

The conductor tips his hat, says "Sorry," and climbs back into the empty passenger car. The train releases a long hiss and then moves slowly past her and down the tracks into the deep bush. She stands on the platform and looks after it a long while, as if a giant hand is pulling, slowly, a stay-stitching thread out of a fuzzy green cloth.

Jack Hodgins

104

River Road

Here I go, once again,
with my suitcase in my hand
I'm running away down river road
And I swear, once again,
That I'm never coming home
I'm chasing my dreams down river road

Mama said, "Listen child,
You're too old to run wild
You're too big to be fishing
With the boys these days."
So I grabbed some clothes and ran
Stole five dollars from the sugar can
Just a twelve-year-old jail breaker
Running away.

Here I go, once again,
With my suitcase in my hand
I'm running away down river road
And I swear, once again,
That I'm never coming home
I'm chasing my dreams down river road

Well I married a pretty good man,
And he tries to understand,
But he knows I've got leaving
On my mind these days.
When I get that urge to run,
I'm just like a kid again
A thirty-year-old jail breaker
Running away

And here I go, once again,
with my suitcase in my hand
I'm running away down river road
And I swear, once again,
That I'm never coming home
In my dreams I still run down river road

Sylvia Fricker Tyson

Stefan

Stefan
aged eleven
looked at the baby and said
When he thinks it must be pure thought
because he hasn't any words yet
and we
proud parents
admiring friends
who had looked at the baby

looked at the baby again.

P.K. Page

The Picture

She had bought the picture at the five and dime store. It was only three dollars, but she thought it was beautiful. Not so much beautiful, as tranquil. That was it. It had a tranquilizing effect on her.

Her apartment was really bare since she had just moved in. The picture was the only thing she had bought to decorate it. She studied it for a few moments. It was a landscape done in very muted colours. Not one thing in the scene evoked any emotion. It had to be the most peaceful scene she had ever noticed in any painting. What attracted her most was the stream, flowing ever so lazily. Perhaps, it wasn't flowing at all.

After work she would lie down to rest and gaze at the picture. She would imagine herself inside of a small rowboat and gently drift down the stream. Her cares and worries evanesced as she projected herself into the picture.

When it was time for her to move, she packed the picture very carefully. Over the next few years she moved several times.

In one house, she asked her sister, whom she had not seen in many years, what she thought of the picture. Her sister's reply had been, "I prefer originals, myself." Well, she thought, that was the difference between her sister and herself anyway. She thought of the dominant picture in her sister's house. It was a ship, painted so as to give a spidery silhouette effect over a bold orange background. It was an original, her sister had paid forty dollars for it.

She noticed no one ever commented on the picture. I suppose most people like something a little more vivid, she thought.

After she was married, and after the birth of her children, she moved three more times. For some reason, the picture lay dormant in her mind for several years. Then, one afternoon, she thought of it again.

She had been lying on her bed, trying to relax, without success. She thought of the picture hanging in another room. It was in an obscure corner and she rarely thought of the picture anymore. As she closed her eyes, she was surprised to find she could visualize it exactly. She thought of the stream and the mossy banks and the huge shade trees under which she would be floating. It seemed so real that as she looked up sleepily she could barely see the sky because the trees formed an arch over the stream. The silent sounds of the forest lulled her to sleep.

The resounding slam of the kitchen door reverberated through the whole house as the children charged in screaming, "Mom, we're home." She was instantly awakened from her lethargic dozing.

The children ran from room to room, yelling, "Mom, where are you?" Finally, they ran to her bedroom. They didn't notice the slight crease on the bedspread. "Mom?" they yelled again as they ran down towards the basement.

She had tried calling to them, but her voice sounded so pitifully far away, they couldn't possibly hear her. The afternoon breeze gently rocked the boat and she reached up to brush away the wisp of hair that had blown across her eyes. Then she dropped her hand to let it trail through the soothing water.

Sandra Merz

The Couch

Cramped between her husband and her father-in-law, Ruby set her jaw and stared fiercely toward town.

"The old couch is good enough," Ansford said. "It's just for sitting."

"It's got the spring through," she replied to her husband before she could stop herself. She had resolved to argue no more. They had argued since Christmas. Arguing was delaying and she was through delaying. Her mind was made up.

"You can sit to one side," Jacob countered. "You don't got to sit on the spring."

On either side, trees rose toward them in a black line, then rushed past with a jerk as though they had been attached to a string that somebody had pulled. Ruby had not been to town for twelve months and the sameness of the road depressed her. She had remembered the countryside as having more variety.

There was no traffic. They had not seen a car since they had left the island. Nor were there any houses or farms. Occasionally, a slash in the thick bush would mark the way to one of the temporary lumber camps that dotted the area. Now that the ground had thawed, most of the camps were abandoned. A few had caretakers, men who had become so bushed they could not bear going to town. When she had first been married to Ansford, Ruby had gone with him to deliver potatoes and fish to a camp. It had been a poor kind of place—a rough clearing dotted with stumps. There had been two bunkhouses on skids, piles of sawdust. There had been a caretaker, a little man with a wooden leg he had carved himself. He wore his pant leg folded back at the knee and he had painted a blue pant leg with a wide cuff, a red sock and a black shoe on the wooden stump. When they had come, he had hidden behind a sawdust pile for five minutes, coming out only when Ansford lifted a mickey of whiskey from the truck and took a drink. After three or four drinks, he had invited them into his caboose. Standing along one wall were a half dozen legs, some with green pants, some with gray, some with galoshes, others with rubber boots. The sight had given her quite a turn. She was not, Ruby had told herself then, going to become like that.

"Aren't you going awfully fast?" she asked as Ansford pulled the truck out of a skid that nearly put them into the ditch. The tie rods were worn and the truck, at the slightest excuse, was inclined to wander.

The roads were barely wide enough for two cars to pass. The edges rose toward the centre to form a convex surface. No matter how much gravel was dumped during the winter, the moment the spring thaw began, the piles of stone sank from sight.

Ansford was too busy wrestling the wheel to reply. Jacob jerked his thumb at the ceiling.

"Going to rain," he said as though she could not see for herself. Clouds lay in overlapping bands like windrows of gray and black stone. "We should go home." Raising his voice to be heard over the gravel drumming like hail on the underside of the truck, he repeated, "Going to rain. Better go home."

Ruby leaned forward, her face composed. She pretended she did not hear him. Ansford glanced at his father but the truck immediately swerved toward the left and he had to haul on the wheel. Although they were not long on the road, his eyes had taken on a wild look and his dark hair stuck out in spikes. He looked as if he'd had a bad fright.

"Go back," her father-in-law intoned, drawing the words out as if he were telling a ghost story. Ruby sniffed. She refused to have anything to do with him unless he had his teeth in. He had a long, lean face that was the duplicate of his son's except that it ended in a collapsed mouth that looked like a shell crater. His false teeth bulged in his shirt pocket. Despite her efforts to get him to wear them regularly, he stubbornly kept them in a sandwich bag and refused to wear them except for special occasions.

Ansford gripped the wheel so hard that the points of his knuckles were white. Before them, the road thrust into the forest like an endless knife.

Ruby secretly studied Ansford's face. Ever since she had first met her father-in-law, she had watched her husband's teeth surreptitiously, finding excuses to look in his mouth. He had already lost two teeth and that worried her. One had infected for no reason. The other had broken while he was cracking hazelnuts. Certain that this was the beginning of a dental disaster, at night, when Ansford was sound asleep, she sometimes rolled back his lip and inspected his teeth as carefully as those of a prize horse.

The three of them had started their trip at five o'clock that morning. She had been so determined to get an early start that she had gone downstairs at four to sit beside the door, her suitcase at her knee. For the trip, she wore rubber boots, slacks, two sweaters and a parka. In her suitcase she had a dress, a hat with two blue plastic daisies, black shoes and a cloth coat. She also had clothes for Ansford and Jacob.

Her husband and father-in-law had come downstairs reluctantly. Ansford came first, Jacob muttering behind him. Normally, both of them sat at the kitchen table but, for the last week, as her determination to buy a new couch had become apparent, they had taken to sitting on the old couch to show her how much wear was still in it. She had not given them an opportunity to delay her but had, on the stroke of five, the agreed-upon departure time, picked up her suitcase and started for the dock. Ansford and Jacob had trailed a couple of yards in her wake.

Neither of them was much good at hurrying at the best of times, but they had hung back like two stone anchors which she had to drag forward with every step. There were no lights in the five houses they passed. She could hear her own breathing and the swish of her boots on the grass. The lake, in this hour before dawn, was flat and dark as asphalt. She climbed down to the skiff, took her place on the middle seat, and sat stiffly, waiting, like a queen about to be transported to another country. Her suitcase pressed against her feet like a deformed child.

The wind they made as they crossed the half mile to the mainland was bitter. The ice had broken up only the week before and rafted ice still stood in piles along the shore like a line of hunchbacked dwarves. As the island faded, the mainland appeared, a solider darkness against the purple sky. Their truck was parked among a dozen vehicles owned by the people who lived on the island. There was no ferry service. Anyone who did not have a boat had to sit and flash his car lights until he was noticed. Then someone came to pick him up.

"We're going on a fool's errand," her father-in-law had said as he climbed into the truck.

When the time came, she thought, burying him would be a chore. While they were carrying the casket to the graveyard, he would still manage to mutter complaints about the inconvenience.

Dawn had been a thin red line enameled to the tops of the trees on the far shore. The line of colour against the sombre darkness of sky, land and water lasted only a few minutes; then thick, solid-looking clouds had filled the gap. In the gloom, lined with forest on both sides, covered over with low clouds so heavy they seemed about to fall, the road might have been the bottom of a trench.

Ruby looked up as a flurry of rain scattered over the windshield. There was a pause which was just long enough to give them hope that the rain would hold off for another hour. Thunder shook the side windows of the truck. Before the noise had faded, rain had poured down. Ahead, the road darkened. Ansford flicked on the windshield wipers. He jerked his foot off the gas pedal and let the truck glide in neutral until they had slowed to twenty.

Neither of them, she thought with a small flush of resentment as she looked at the mud, ever remembered to take off his shoes at the door. It was a constant source of aggravation. No matter how often she washed the floors, they were never clean. Both men left trails of mud wherever they went. She had tried putting a mat outside the door. When that had not worked, she had added a sign saying "Boots here". She had spread newspaper inside the door. Finally, when it rained, she had taken to making a path of newspaper from the back door to every room in the house. The first time she had done it, they had carefully side-stepped the paper, certain it was for something special.

Ansford and she had met at an upgrading class in Winnipeg. The government was paying them both to attend so that they would not be counted as unemployed. For Ansford, squeezed between a bad fishing season and the need for new nets, it was a chance to get cash money. For Ruby, it had been a chance to get off her feet. She had been a cashier at Safeway for five years. Nine months of the year, a cold wind blew across the floor every time a customer came in or went out. The manager, thinking it would keep the customers' minds off rising grocery prices, insisted that the girls wear short skirts and shoes instead of boots. The constant standing on concrete had covered her calves with a fine mesh of red veins.

Ansford had brought her to the island late one night. Douglas McBrie had taken them across. He had said nothing to either of them but as she was climbing out of the boat, Ruby heard Ansford say, "I got her and a new outboard while I was in the city."

Ansford had told her he had a house and furniture. Her father-in-law had come as a surprise. When they arrived, the house was dark and silent. She and Ansford, tired from a long day on the road, had gone straight to bed. He had fallen asleep but she, overtired and excited by the unfamiliarity of the house that was to be her home, had been unable to sleep. Restless and not wanting to wake Ansford, she had started to go downstairs. Jacob had been standing in his nightshirt in the hallway, a tall, thin ghost with knock knees. She had been wearing a blue flannel nightgown and had her hair set with pink plastic curlers. Jacob and she had stood rooted in place. The look that crept into his face as the shock passed made her feel like the whore of Babylon. Wordlessly, she had fled, shutting the door behind her and dragging a chest of drawers against it. Ansford snored gently.

The next morning, Ansford explained to Jacob that he had married. Jacob had nodded but his eyes and the pensive set of his mouth revealed that he did not really believe it. Even the marriage certificate did not erase his lingering doubt. She and Ansford had been married twelve years, but Ruby was sure that Jacob, somewhere deep behind his eyes, believed they were living in sin.

The truck began to jerk. Ruby looked up. "What's the matter?" she asked sharply.

"Mud," Jacob yelled. His hearing was not good. To hear his own voice, he had to shout.

Ansford was pressed so close to the steering wheel that he seemed to be impaled upon it. His light brown hair looked nearly mahogany in the gloom.

Ruby braced her feet on the floor and raised herself from the seat so that she could see herself in the mirror. She had carefully set her hair the night before but already the violent motion of the truck was making it untidy. Swaying with the truck, she used her fingers to comb her hair into place. She prided herself on not letting her standards go. Many of the women did. They looked fine but the moment they got married it was as though some tightly twisted rubber band inside them was snipped and they began to fall apart. Every Friday, she religiously set her hair and changed the bed sheets. Some women she knew never combed their hair, never mind set it, and changed the linens every spring. She had told Ansford right after they married that she could not live like that.

Jacob guarded his sheets as if they were jewels. At first, she had fought with him but, gradually, she had learned to have her way without conflict. She simply waited until he left the house and, since none of the doors, including the outer ones, had locks, she went into his room, stripped and remade his bed. Occasionally, if his liver was bothering him, he still berated her, beating his fist on the table, shouting, "You leave my sheets alone, you hear! They're fine. You'll just wear them out." Of late, she had noticed that his voice was no longer sharpened by conviction.

Instructed by the country memories of parents who had moved to the city when she was only two, she had come prepared for the wrong battle. Life on the island was not endless visiting in other people's kitchens. Work seldom ceased. Leisure was idleness forced upon people by blizzards. The men cut, repaired, dug, ploughed and fished endlessly.

Their wives, inundated with children, never gave over cooking, washing and mending.

Ansford geared down to second. The truck, like a cow that has tried to dodge one way, then another, and failed to escape the drover's switch, settled into a steady run. Ruby opened her purse. It was the shape and size of a shopping bag and made of purple plastic. She took out a round mirror so she could check her makeup. Her face was broad as a pumpkin. Although she constantly dieted, she was still heavy. Hard work had settled her flesh downward like buckshot in a cloth bag.

Jacob's left arm was flung along the back of the seat. He gripped the metal rail fiercely. His long legs were stretched out and his right foot jerked up and down every time the truck began to skid. His face was strained.

Ruby leaned toward him and held the mirror in front of his face. "Put in your teeth," she yelled in his ear. "You don't want to get killed looking like that."

He jerked his head around like a hawk and tried to stare down his nose at her but he could not keep his eyes off the road. He could stand in an open boat five miles from land with waves twenty feet tall flinging him about like a cork while he picked fish out of his nets, but he left the island so seldom that each journey took all his courage. Having to ride over bad roads made it worse.

"We got to go back," he shouted. "It ain't worth it to have a new couch."

Out of the corner of her eye, Ruby saw Ansford sneak a glance at her. She knew that if she showed even a moment's weakness, he would turn around and race back home.

"Keep going," she ordered. She was glad she was two years older than he. If she had not been, she would not have had the sense to know that he needed to be told what to do.

"I can't take much more of this," Ansford said. He was holding the wheel so hard that he looked as if he were going to pull it loose. He did not drive much and when he did, he preferred to travel in the centre of the road at thirty miles an hour. At such times, he sat as far back on the seat as he could, his chest thrown out and his head tipped up so that he was just seeing under the sun visor.

"I got to stop," he said, his voice a defeated whine. Ahead, a driveway made a shallow loop nearly parallel with the road. A tall gas pump with a round glass top like a fish bowl sat in the centre of the driveway. On the edge of the bush, a low green building squatted close to the ground. Ansford parked beside a black Chevrolet with orange ball fringe on all the windows.

They stopped at the door and, leaning in, their heads pressed together like balls, studied the interior. A naked light bulb burned over a single pool table. A group of eight Indians were frozen into position around the green felt, their eyes not looking anywhere, their ears listening to every sound. Ansford led the way to a counter with four stools.

"Bad day to be out," a fat man in tweed pants and a white shirt said. On the pocket someone had embroidered *Jimmy*. The left corner of his mouth and his left eyelid drooped. The last part of each word he said was slurred.

"Whad he say?" Jacob demanded, rifling his pockets for change.

"He said it was a bad day out," Ansford said. Ruby was sitting between them so Ansford had to lean steeply to one side to get his mouth close to his father's ear.

The Indians had started to play pool again, but when Ansford shouted, they froze into place, shining under the light like pieces of old walnut furniture.

"A bad day," Jacob repeated, nodding to himself, his eyes reflective.

He turned to Ruby, thrust his face close to hers and bellowed. "It's a bad day for being out. We should turn around and go home."

"Coffee," Ruby said. "Black."

"Whad she say," Jacob asked Ansford.

The Indians still had not moved.

"Coffee," Ansford replied at the top of his voice. Jacob nodded vigorously. "Me, too."

Jimmy brought their order for them, dropping the heavy white crockery with a clatter.

"You must be Jimmy," Ruby said.

The fat man shook his head. "I'm Bill."

"Whad he say?" Jacob asked.

"My name's Bill," the fat man shouted.

They all studied his pocket. He looked down.

"Got these shirts at an auction," he explained, raising his voice as if he were talking to a multitude. "They all got different names on them."

"It says Jimmy," Jacob insisted.

Before Bill could explain again, Ruby said, "I don't remember you from last year."

"Christmas," he replied. "Under new ownership. Had a sign. Wind blew it down." He shouted each phrase so that Jacob could hear.

"You from the city?" Ruby asked. He nodded. "I seen your white shirt and I knew." She shot a glance full of reproval at Ansford. "My husband's got a white shirt but he won't wear it but once a year."

Ansford leaned closer to his coffee. He had

spread his elbows on the counter and was counting the different kinds of candy bars. There were five kinds he had not eaten.

"I'd give a dollar to go back home right now," Jacob declared.

"You'd think we made him come." Ruby pressed her lips together in mild indignation the way she might with a child. "He complains all the time but if we leave him alone for half an hour, he comes looking for us."

"The road's bad further on," Bill said.

"You seen these?" Jacob asked. He dug into his jacket pocket and held out a rooster carved from a forked stick. "I sold lots of these. Dollar apiece."

Bill saw Ruby looking at the name on his pocket. "Tomorrow, I'll be Norman. The day after that Robert."

"If you bought five for a dollar," Jacob said, peering at Bill's face, trying to see if there was a flicker of interest, "you could sell them for a dollar seventy-five."

"How do you know who you are?" Ansford burst out. "Every day you've got a different name."

"Who," Bill said, turning to look square into his face, "ever knows who he is?"

Jacob was holding the rooster between his thumb and forefinger, twirling it around so that all its good points would be revealed. Bill turned back and nearly got the rooster in the eye.

"He likes to whittle," Ruby said. "Finish up," she said to Ansford. Ansford was trying to make his coffee last as long as possible.

Bill reached a large hand under the counter and drew out a brown envelope. "Would you deliver this to the garage? It's a cheque for some car parts. They won't send any more until they get paid."

Ruby slid off the stool. "Not much business here." She took the envelope, and squashed it into her jacket pocket.

"Suits me." He folded his hands over his stomach. "Had a stroke. Can't do more than a little."

"I wish you luck." She said it with solemnity the way she might to someone who had declared he was going to jump off a cliff and try to fly.

"I got a pension."

Ruby adjusted her head scarf and started for the door. Behind her, Bill called, "It ain't much but everybody's got to have something to keep him going."

They hurried through the rain that cut across the sky. The drops fell with such force that the water might have collected around a centre of lead. Inside the cab, Jacob said petulantly, "If you'd a just waited, I'd a sold him some carving." When he felt unjustly treated he had a way of drawing his eyebrows together until they nearly touched.

The road was as slick as if it had been greased. Rain fell steadily. The ditches were full. Ruts had become long bands of nearly black water.

Jacob sat deep in thought for the first mile and she thought he was sulking, but all at once he straightened up, looked at them both and said, "He must be hard of hearing. He certainly shouts a lot."

They traveled the next thirty miles in silence. Because of the rain, there was nothing to see and Ruby and Jacob fell into a light doze until Ansford startled them by sharply calling out a warning.

They were approaching a bog. The land on either side of the road was a sea of moss. Full-grown trees were no more than three feet high. When they crossed the edge of the bog, it was as if the truck had been grabbed from behind. Their bodies were flung forward. Ansford gunned the motor, accelerating as fast as he could. The truck jerked so violently that it felt as if the transmission had been ripped out. Ansford slammed the truck into second, then first. It was no use. The truck slowed; the motor coughed and died. Ruby could feel the wheels sinking.

They sat, staring into the rain as if hypnotized. There was nothing before them except empty road and endless forest. On either side of them there was forest. Behind them, blurred by rain, the trees went on until the road disappeared and the two dark lines converged.

"We can't stay here," she said.

"We're twenty miles from town," Ansford protested.

"We can't go back. We've come too far."

Ansford looked out the rear window. When he turned back, the skin on his face was tight as though someone was pulling it from behind.

"We should of stayed home," Jacob wailed. "I told you. Every year it's the same. The ice goes off the lake and she's got to go to town for something."

"Let's go," she said, her voice determined. "Sitting here talking isn't going to get us anywhere."

She gave Ansford a shove with her hip.

"I'm not going," Jacob screamed. "I'm an old man. My legs won't take it."

Ruby gave Ansford another shove. He opened his door and got out. He lifted Ruby's suitcase out of the wooden box on the back of the truck. As they started away, Jacob yelled, once more, "I'm not going. I don't have to go just because you say so."

The rain beat on them as they ploughed through the mud. Mud clung to their boots like paste. Mud splashed up their legs. With every step, Ruby had to pull her foot loose. The ground was a quagmire. Each time she lifted her feet, more mud clung to them. After fifteen minutes her feet were so grotesquely large that she had to stop. Behind her, she heard water splash. She turned around. Jacob was standing nearly on her heels. His shoulders were hunched together. Water streamed through his thin hair and poured in a steady stream from his nose. Bending over to protect it from the rain, she reached into her purse and took out a second rain cap of clear plastic. She pulled the cap over Jacob's head and tied the bow beneath his chin.

Seeing his dismay, she shouted, "Nobody's going to see you."

Ansford came up to them with sticks of willow which he had cut. He had flattened one end. With these, they pried the mud from their boots.

The rain swept down, engulfing them, breaking over their heads and shoulders like surf.

Ruby tried various strategies. She tried to pick her way carefully, placing her feet where the gravel was thickest. All that happened was that both mud and gravel came up together. She tried walking on the high spots on the theory that the ground would not be so wet. Then, in the hope that the water would wash her boots clean, she walked in the ruts. Nothing worked. After a while, the weight of mud on each foot was so great that she had to swing her legs stiffly from the hips.

When they reached what she had estimated was the second mile, she checked her watch. It was fifteen after twelve and they had eighteen miles to go. She looked back. The truck had disappeared behind a curtain of rain. In front of them was an endless stretch of mud and sodden trees. All she wanted was to reach the town. The thought that they still had eighteen miles to go made her waver. At the moment, if she could have, she might have agreed to return home. Since that was impossible, she forced herself ahead by thinking about lying in the bed at the hotel and watching TV.

At one-thirty they had to stop to rest. They had walked the last half mile on rising ground and the going had been easier. Now, the ground sloped down again. As she stood, looking along the road, Ruby's thigh muscles felt as if they had been pulled loose. Her knees felt as if the joints had been ground down with pumice. Water was seeping through the seams of her jacket.

"We've got to keep going," Ansford said. Shorter and broader than his father, he still looked lean and hard. He was slightly bandy-legged, and his wet trousers clinging to him emphasized the two outward curves. Water dripped from his plaid cap. He took it off and wrung it out.

He was slow to make up his mind, but once he had there was no stopping him.

They started off again. Lifting her feet hurt so much that Ruby thought of taking off her boots and socks. Less mud, she was sure, would cling to her skin. She rejected the idea because the mud, only recently thawed, was still cold. Her breathing was beginning to be laboured, her breath whistling in her head. She shivered and wished that she had something hot to drink.

Darkness settled over them so gradually that she was not aware of the fading light until she began to find it difficult to see Ansford. Earlier, she had been worried about being caught on the road in the dark. Now, she was too tired to care.

The rain still fell, no longer in torrents, but in a steady, chilling drizzle. Her legs were soaked. Her face was numb.

The walking had turned to stumbling. There were pauses between steps. At last, they stopped and huddled together.

"How far do you think it is?" she asked.

"Ten miles," Ansford replied, "maybe eleven."

"We've walked all day," Jacob cried, his voice thin as a spider web.

They staggered forward for another hour. Ruby's feet, heavy as cast iron, dragged through the mud. When she realized that Jacob no longer was behind her, she called Ansford. They started back. They found Jacob sitting in a rut, struggling feebly to get up, Ansford took one arm, Ruby the other. Between them, they heaved him to his feet. Since he could go no farther, they led him to the edge of the forest. Even here, under a canopy of branches, rain sifted down upon them. They stood dumbly, unable to see, too tired to want to do anything except lie down and rest.

They turned Jacob in a half-circle, positioned him between two saplings rising from a single root. They pushed him down, cramming him between the two trunks so that he was firmly held in place. Ruby sat on the left, Ansford on the right. Oblivious of the rain or their aching bodies, they fell asleep.

Ruby woke up cold. She could not feel anything from the waist down and, at first, she was not sure where she was. Her legs were stretched out before her like two dead weights pinning her to the sodden ground. She lifted her left arm to pluck at her parka and tried to pull it more tightly about her. The clouds might have been molded from clay. She was, she realized, still leaning against Jacob. His chin rested on his chest. She knew he was still

alive because she could see his nose dilate with each breath. Ansford had fallen over and lay on his back, his mouth open. She wondered, her thoughts distant, detached, how it was that he had not drowned. Mercifully, the rain was only a fine mist.

Grasping the tree with one hand, she pulled herself to her knees. Gradually, as she kneaded her legs, the blood came back into them. She dragged herself to her feet. She did not dare let go of the tree. Her suitcase, she noticed, lay beside Ansford's hand. She was glad that she had wrapped all their belongings in plastic.

She reached out with her toe and jabbed Ansford. He did not move. She kicked harder, digging her toe into his ribs. He opened his eyes, lay staring at the clouds, then closed his eyes again. She kicked him hard enough to hurt her big toe.

"Get up," she said. Her throat was so cold that the words were a croak. "You've got to get up." She had never quit shivering and every time she shut her eyes she had the sensation of falling. She knew that she could do nothing without him.

He opened his eyes, coughed twice, rolled over and pushed himself up. He looked as if he had been dug out of a grave. He was covered from head to foot in mud. Pressed into the mud were twigs, leaves, grass, pine needles, even a couple of feathers. His hair was matted, his eyes sunken.

"Jacob," Ruby said. She slapped him on the back. There was a wet smack. He groaned. She hit him twice more. He stared dumbly, his eyes unfocused. They each took an arm and because they were weak with hunger and cold had to strain to pull him free. At first, he was a dead weight. They walked him in a tight circle, around and around in the wet grass. His legs kept collapsing. One moment, his legs held him, then they gave out and he dropped to his knees. Each time, grunting, one hand under each armpit, the other braced just above his elbow, they levered him up. It was like walking a horse with severe colic.

"We'd better get going," Ansford said. He picked up the suitcase.

The day before, mud had clung to their boots. Now, the mud was so wet that it was the consistency of tomato soup. With each step, Ruby sank past her ankles. She could think of nothing except being dry and warm and eating platters of food. Gradually, walking warmed her, but her hunger grew into a savage pain. Even that, however, passed into a dull ache. She felt as though she had swallowed a large, smooth stone.

It was noon when they heard a noise like distant thunder. At first, they ignored it. They staggered forward, their bodies lurching from side to side.

Lifting her eyes from the mud, Ruby looked past the rounded hump of Ansford's back. Moving slowly toward them was a tractor. Ansford looked up and stopped. She saw his shoulders settle as though air had been holding them up and had suddenly been released. She thought he might fall down but, instead, he stood wedged in the mud like a fence post. Ruby stopped, grateful not to have to lift her feet again. She felt that if someone touched her with the tip of his finger she would topple to the ground and be unable to rise. She could hear Jacob splashing behind her. When he came abreast, she caught his arm sharply.

He had been walking automatically, his body moving independently of any thought. Her fingers on his jacket sleeve stopped him as completely as if someone had turned off a switch. He stood and trembled, docile as a tamed animal. The tractor churned toward them.

The tractor driver's back was covered in mud kicked up by the chains. He grinned at them, then got down and gave each of them a hand up.

"Bill at halfway house phoned to say you were bringing some money he owed me. When you didn't arrive, I figured I'd better come looking for you."

"Spent the night in the bush," Ansford replied.

"Looks like it," the driver said and turned the tractor around.

There was no place to sit so they rode standing, Ruby and Jacob on either side of the driver, Ansford on the hitch. They climbed down in the middle of Main Street. The rain had melted the mud just enough to spread it evenly over them. Jacob still had on Ruby's head cover.

The tractor driver promised to bring in the truck; then Ruby led the way to the hotel. They rented a room with a double bed and had a cot put in for Jacob.

Ruby had a hot shower with her clothes on. When the water running along the bottom of the tub was no longer gray, she peeled off her parka, waited for the water to clear again, then undressed completely. She washed, changed into the clothes she had brought, then helped Jacob into the bathroom. He moved stiffly. He tried to undo his jacket and could not, so she stood him in the tub and turned on the shower. She undid his parka. The warm water started to revive him. She helped him off with his sweater, undid his shirt and pulled it off. She threw everything into the tub to be washed later. He let her pull off his socks but when she started to undo his pants, he protested. She told him to be quiet or she'd call Ansford to come and hold him still.

When he stood in nothing but his long underwear, she gave him the soap and left.

Ansford lay curled before the door on a wad of old newspapers which the desk clerk had given them. All the time he waited, he shivered and jerked. Except for his eyes, he was completely caked in mud and, as it dried, it stiffened so that he looked like an unfinished statue.

Ruby laid out their clothes. She did not have much room in the suitcase but she managed to bring a change of clothes for each of them, including a white shirt for both men.

While Ansford was washing, she took their dirty clothes to the basement of the hotel and washed and dried them. When she went back upstairs, Ansford and Jacob looked scrubbed and brushed. The three of them went to the dining room and ate two platters of ham and eggs each and drank twelve cups of coffee between them. By the time they were finished, they all had a satisfied, glazed look.

Ruby led them down the sidewalk to the Red and White Hardware and Furniture store. A salesman in a brown suit scurried out from behind a small forest of pole lamps. Ruby scanned the room. The salesman, his eyes full of anticipation, his hands washing themselves in little circular motions, darted this way and that.

"I heard you've got a blue couch for sale," Ruby said. "We need a new one."

The salesman lifted himself up on his toes as though he was a ballet dancer and, wobbling, looked across the array of couches. "I've got a nice red one," he said.

"You've got a blue one," Ruby insisted. "With pansies. I heard at Christmas you had it."

"I sold that months ago," he said.

"I wanted blue," Ruby insisted.

The salesman asked her to wait, hurried away and came back in a minute with a catalogue. He pointed out a picture of a blue chesterfield and offered to order it.

"It's not the way I imagined it." Ruby pursed her lips in disapproval.

"I got others," the salesman said.

"No," Ruby answered. "It isn't what I thought." Her decision made, she turned around and herded the two men before her.

"I told you," Jacob whispered indignantly to Ansford. He moved his teeth about with his tongue. "I told you she wouldn't buy it. Every year it's the same." His voice carried all the way to the other side of the street.

Ruby ignored her father-in-law and stood, her hands on her hips, looking with satisfaction up and down Centre Street. It was only a block long but the stores were deep and contained a host of objects. There were women in town whom she knew well enough to visit and on Sunday there was a service at the church.

The rain had stopped. The clouds were breaking up to reveal a clear, bright sky. The street, covered with a thin layer of water, looked as if it had been plated with silver. The signs of the stores, swept clean by rain, were bright and shiny.

"All that," Jacob complained, "for nothing." But there was no force in his voice, for he was squinting, trying to read a sign on the movie house a half block away. In any case, Ansford was paying him no attention. He had turned halfway around to try to identify the half dozen men who were sitting in the window of the garage.

"What can't be helped shouldn't be mourned," Ruby said, seeing her reflection in a puddle as shiny as a newly minted silver dollar. She tilted her hat so that the flowers showed to more advantage. "We can't," she added, "go anywhere until the road dries. There's today and tomorrow. We'll just have to make the best of it."

W.D. Valgardson

Letter from a Small Island

On a sunny Saturday in May of 1973, Earl L. died. He was sixty-one then, and a robust man with rosy cheeks and cheerful blue eyes. I had known him for about a year then, and knew him as a friendly man who had a smile and a bright greeting and a little joke for everybody he met in the course of the day.

He was also a discreet man. He kept his opinions and plans to himself. His last years were those of "Resettlement"—the evacuation of hundreds of small fishing communities—and whereas the island had been home to some eighty families in 1967, the population had declined to seven families by 1972 and to three families by 1973. But when Earl was asked whether and when he would "shift to the main" he would never commit himself: "Now that's hard to say. . . . "

He lived in an old two storey house that was very well kept up. His wife was an exceedingly shy woman who rarely left the house. His only child, a daughter, had long ago moved to the city.

Earl was the skipper of a small "longliner". He went out on it with a teenage boy to set a fleet of cod nets that Saturday, early in the morning. I met him on the road just before he left, and we exchanged a few friendly remarks. He looked hale and hearty.

I went over to the mainland later that day to go dancing in the evening at the Lion's Hall in Arnold's Cove. I have good friends there; it was my habit to spend a weekend there about once a month.

At the dance, I was sitting with Ann and Wilson and two of their nine children. Toward eleven o'clock, when I came back again to the table from the floor, Ann asked me with sudden gravity: "Guess what happened?"

I thought that one of her children had become involved in some trouble. I said: "I have no idea. What?"

"Earl died up on the island about two hours ago! Sheila just phoned to let us know."

Sheila is Earl's niece. She had worked for a couple of years in the city as a nursing assistant, but had come back to the island to marry a young fisherman there.

"Why, I just talked to him this morning, and he looked as healthy as—was it an accident?"

"No. He had a heart attack. He ate a good supper and then watched TV. Around eight o'clock he said he wasn't feeling so good, and went upstairs to lie down. A while later he got up to go to the bathroom, and his wife heard him fall in there."

I asked: "Is the doctor going up to the island to have a look at the body, or are they going to bring Earl to the mainland for an autopsy?"

Ann shook her head a bit bewildered: "No. Sheila told the doctor over the telephone what happened, and he is going to mail the death certificate."

I was amazed when I heard that. I had lived for years in the city and had become accustomed to look at death as upon a condition which is almost shameful and which requires that bodies are whisked away and cut open and inspected and disembowelled and painted and then presented to the bereaved as undead as possible. It shocked me that death should be accepted so readily here, that no vengeance should be wrought upon Earl for reminding us with his death that we are all mortal.

But then another thought struck me: how marvellous to live on an island where foul play was completely unthinkable in connection with death!

On Sunday, when I returned to the island, the small harbour was full of boats of all sizes which had brought a hundred or more of Earl's former neighbours and friends to the island. They came to express their sympathies to the widow, and to then linger at a lilac bush or a few concrete steps or a crabapple tree where their homes had stood only a few years previously.

Some of Earl's friends washed the body and dressed it in Earl's best outfit.

That May was unusually hot. It was not until Tuesday that the coffin arrived and an even larger crowd of people gathered for the funeral. I have been told that the body had by then begun noticeably to decompose, that the friends who put it into the coffin came downstairs again with measured steps and walked gravely through the kitchen and took pains not to slam the door while they left the house through the back porch, and that they then raced uphill to vomit in the woods there.

Since I was neither an old acquaintance, nor a relative, I was asked to help in digging the grave in a cemetery which overlooks the sea. While Earl was brought to the little white clapboard church about a mile from the settlement in an old blue van, six of us took turns on picks and shovels— some were visibly uneasy and fortified themselves with strong liquid spirits.

At one point, the side of the new grave fell in and revealed the rotten boards of a long buried coffin. We all waited for an anxious moment apprehending that the board might give way and spill a rotten corpse or a skeleton.

The ground was wet, and no sooner had we dug the grave than the water started to rise in it. We bailed until the funeral procession entered the graveyard.

In front, the pallbearers were redfaced and perspiring. Behind them, the widow swooned with grief, the daughter was pale and rigid, Earl's brother looked grim, the latter's wife seemed ill, and their children were visibly uncomfortable. But a few rows further back the men were already talking about fish and boats, and toward the end of the long column young people were shoving each other off the path into the spruce trees.

The coffin was lowered. The widow groaned. The clergyman opened his book and began to read.

But the water rose rapidly in the narrow space around the coffin. Before long one end of it seemed to stir. The clergyman, perceiving the situation, read the graveside service with increasing speed until he sounded like a recording at wrong speed.

The outcome was in doubt for a few instants. But at the earliest possible moment soil was thrown on the coffin with almost indecent haste, and it did not rise anymore.

Yesterday I visited Earl's grave for the first time in three years. The fence around the graveyard has fallen down, the forest is encroaching. Earl's grave is already covered with grasses and wild flowers. Three little spruce trees grow on it.

I am forty-one now, at an age where one slowly accepts that one will have to die eventually. And may my friends forgive me! I think I want to die like Earl. I want to leave this world at the end of a sunny day full of good work, rather than be sickly and bored and angry at the approach of death. I want to have friends who wash and dress my body and put it into the coffin, rather than have it inspected and mutilated and decorated by doctors and morticians. I want to be buried where winds and waves and birds are the usual sounds, and where flowers and trees slowly obscure my grave and ease me out of the memory of friends and family.

Randy Lieb

St. Leonard's Revisited

We came ashore
where wildflower hills
tilted to the tide
and walked
sad and gay
among the turnip cellars
tripping over the cremated
foundations
of long-ago homes
half buried
in the long years' grass

almost reverently
we walked among the rocks
of the holy church
and worshipped roses
in the dead yard
and came again to the cove
as they did after rosary
in the green and salty days

and men offshore
hauling traps
wondered what ghosts
we were
walking with the forgotten sheep
over the foothigh grass paths
that led
like trapdoors
to a past
they could hardly recall

Al Pittman

Death of an Outport

In the summer of 1967 I sat in the kitchen of a fisherman's home on the island of Merasheen in Newfoundland's Placentia Bay. The fisherman, Anthony Wilson, had seen my wife and me walking down the road past his bungalow and, because we were strangers, he had invited us in for a cup of tea.

In Newfoundland "tea" means a fully laid table including linen cloth, the best china in the house, home made bread, a variety of wild berry jams, a platter of luncheon meat and always a jar of molasses. After we had gorged ourselves on Mrs. Wilson's "tea", Anthony broke out a bottle of rum. He had had the rum come in by mail boat two weeks before and had ever since kept it hidden away in the bedroom only to be opened on the day of the annual garden party three days hence. Anthony, however, decided that having strangers in was excuse enough to break the rule and promptly produced a bottle of black demerara. For an hour we passed the bottle back and forth across the width of the kitchen table and talked of Merasheen.

Merasheen lies about five miles off the west shore of Placentia Bay on Newfoundland's southeast coast. Most of the island's inhabitants live on the island's southern end in the villages of Merasheen, Little Merasheen, and Hickey's Bottom. The villages are located in three adjacent harbours, affording the fishermen of the place a choice of landings when weather conditions prevent them from going in to their usual moorings. Behind the villages lie the barren sheep-dotted hills of Merasheen which give the island its bleak, naked appearance. Beyond the hills, however, there are miles of forest where the men snare rabbits in the fall, and beyond that, more miles of barrens where caribou roam out of range of the guns of the American big game hunters who come in droves to Newfoundland each autumn. Though the people of Merasheen feast on rabbit stew and caribou steaks in season, their livelihood is harvested out of the dark Atlantic waters that are everywhere around them. The violent rhythm of the sea is the rhythm in which the people of Merasheen have lived since man first set foot on the island's rugged perimeter.

My father was born in Merasheen in 1907 and I was born thirty-three years later in the tiny village of St. Leonard's just across the bay. I had gone there that summer with my wife to put all the stories my father and mother had ever told me into

their proper setting. I had been taken out of the bay before I was six months old, and though I knew Chapel Pond and the Jawbones, and Soldier's Point, and the Jigging Cove and St. Kyran's like the palm of my hand, I had never seen any of them. So I went that summer to see where my father had come from, where my mother had come from, and where, most of all, I had come from.

"Sounds like a hard way to make a living," I said when Anthony had finished telling of one particularly rough time he'd had in winter fishing.

"Well, I'll tell ye Phonse," he replied, "it's the devil's own handiwork betimes, but once ye leave off on a summer morning, heading out, with the sun just peeping up, the skiff cutting clean in the water, and all that shiny sea stretching out ahead of ye to westward, well Phonse, you go out one morning like that and you can put up with winter fishing the rest of your life."

It wasn't at all the sort of thing I would have expected from the weather-beaten, granite giant of a man sitting across the table from me. Yet, when he said it, it rang so true I felt a sudden surge of sadness sweep over the room, for as we sat talking, we, all of us would head out from Merasheen.

Centralization, Premier Joseph Smallwood's plan to "drag Newfoundland kicking and screaming into the twentieth century", had already taken its toll in Placentia Bay, St. Leonard's, St. Kyran's, Clattice Harbour, St. Anne's, Toslow, and numerous other villages were already being reclaimed by the wilderness into which they had been etched some hundreds of years ago.

The livyers had been paid a subsidy to move to a better life in places like Marystown and Placentia, where, they were promised, there would be jobs galore, and motor cars, and television sets, and better educational facilities for their children.

If such well-timed persuasions as these failed to move the people, the church lent a helping hand. It closed down schools and churches and took away the priests. Inevitably the latter did the trick. The people of Placentia Bay outports could do without cars, and supermarkets, and television sets, but being as religious and as superstitious as they happen to be, they could no more think of living where there was no priest than they could think of living inland. So they moved.

From all the villages of the bay they moved to the government designated "growth centres" where they discovered, too late, that the only growth was the growth in population—the result of their own mass migration. Too often they found the worth of their subsidy not nearly enough to

replace the homes they had left behind in the coves and on the islands. Too often they found that the promised jobs were nonexistent. In Placentia, for instance, where so many of them were sent, they found that houses were hard to come by, and jobs even harder. The only sources of employment in the town were the Canadian National coastal boat terminal and the American naval station at Argentia. Due to cutbacks in CN coastal service (now that there were fewer outports to serve) and to cutbacks in U.S. military commitments in Newfoundland, there were perhaps fewer employment opportunities than ever before.

And now there was talk of Merasheen.

It seemed there was nothing Anthony Wilson or anyone else could do about it. The government fish plant was closing down, therefore there'd be no market for their fish. They could, as they did for years before the fish plant opened, take their catch to Wareham's in Harbour Buffett. But Wareham's too were curtailing operations.

As well as closing down the fish plant, the government would also halt operation of the dynamoes that had, for the past few years, delivered electricity to the islanders' homes. The school had already closed. And the priest was leaving in the fall. So the people of Merasheen would have to move. What else could they do?

Anthony Wilson didn't want to go. He had his own home, and a comfortable and sturdy dwelling it was too. He had a garden out back where his wife grew turnips, potatoes, carrots, beets, cabbage, and a variety of currants and gooseberries. And when I suggested that his fishing take would probably do no more than pay for the gear, he said, "No, Phonse me son, we does a bit better than that."

And when his wife went to the bedroom and returned with the new clothes she had bought by mail order for the children and herself so that they might look "fine" on the day of the garden party, it wasn't hard to tell that the pleasure of the newly acquired finery was in no way diminished by the thought of payments, installments, or "time" as Newfoundlanders refer to credit.

But they would go. There was no other way.

Mike Casey would go too, and his wife Elizabeth, though she kept saying over and over that they would have to drag her away.

Stan Ennis and his son Andrew would go too, though they owned one of the best boats in the bay and Andrew was as good a fish-killer as his father. And George Wilson would go. Skipper George Wilson, white haired, as tall and dignified as a

church spire, skin the texture of rawhide, bread 'n' buttered there some eighty odd years ago, a legend in his time, father to Anthony, village elder, as gentle as the waves lapping the shore below his house, as rough as the rock that threw the sea back upon itself when it erupted with all its fury upon the Jawbones. He would go too. Go leaving his wife's grave to the delinquent sheep. Leaving all he would have passed on to his sons to the wind and the rain and the sea. Would go leaving everything behind with his memory and his old man's heart. But he would go.

"I could see it," Anthony said, passing the bottle, "I could see it maybe if they moved us all into St. Kyran's or any place down here in the bay. There's good harbours, the fish is here, the men is here what can catch 'em too. I can see they wants bigger schools. I understands that. We been having hard enough time getting a teacher to come here and they only stays a year at the most. I can see the priest wanting one church to look after 'stead of a whole bunch of 'em. God knows, he has it hard going at it all the time. And a lot of priests don't like it in the bay no more. Well, they isn't

fishermen so's I don't know ye could blame 'em any. All the same though, I can't for the life o' me see why they shifts us to Placentia. Ye knows yourself there's no living to be made there. The base is closing down bit by bit. Where's the men going to work, I asks. A man can't fish outa Placentia, that's for certain and for sure."

I took a long swig on the bottle and regretted that we couldn't stay for the garden party on Sunday. If I had my time back now, I would have stayed no matter what. But at the time the significance of it all passed me by. It didn't strike me as it should have that this garden party would be the last ever to be held in Merasheen. It would be the end of a tradition that went back before my father's father's time. The end of a way of life.

The morning after our visit with Anthony we walked past the parish hall and saw the tarpaulin booths all in a row in the church yard. Sunday they would be ringed by little girls in floral print dresses; by the men of the place, coat pockets bulging with bottles, Sunday tweed caps angled on their heads; by women with babies on their hips, white aprons looking altogether fine in the outdoors; by young girls with the dishes already done and for the first time in three days no rollers in their hair, flirting openly with the Peters and Andrews and Jims of Merasheen; by the boys who made root beer from extract and carried it in bottles, as drunk as their fathers in their fantasies.

Sunday the booths would house ice-cream in heavy canvas khaki bags, wheels of fortune, cabbage-roll dinners, ticket pedlars, bean bags, balloons and darts, sacks for the sack race, ropes for the three-legged race, steaming boilers of good things to eat, coca cola in cases, peanut butter kisses, licorice, and home-knit scarves and caps and socks and mitts to be won as prizes.

But that morning the booths stood empty, their sides flapping noisily in the wind, as they had on that same morning for hundreds of years past.

The Devil was there too, looking very much out of place in the middle of the empty yard. On Sunday every man and boy in Merasheen would take a crack at knocking his head off. How long ago was it that some expert young chucker first knocked the Devil's head off, sent it rolling beneath the feet of the crowd, heralding good tidings for the people of Merasheen?

We left the parish ground and went over the hill into Hickey's Bottom. Mike Casey came then and invited us "'ome to 'ave a shave and to meet the missus". As we walked along the beach road, Mike pointed out to me the precise spot where my grandfather's house used to be and the path he used to take "luggin' 'is long tom" going into the barrens to get rabbits.

I could see my father, a little boy, running up the path at dusk to greet him, tall like timber, coming home from a day's hunting on the barrens with his long tom over his shoulder and a brace of rabbits dangling at his side. A vigorous man, still vigorous after a day's trek on the barrens, tossing his young son high into the air and carrying him secure on his shoulder to the house.

"I'm too old to be going anywhere's at my age," Mike said as we sat in his kitchen nipping on his garden party rum. It seemed the invitation to shave was just an excuse to bring strangers home so that he could get at the rum without his wife objecting.

"What the jeesus ye expect a man o' my age to be doing in Placentia, I asks. Lived right here all me born days. Ain't no time to be gallivantin' around at my age."

So he talked on through half the bottle of dark rum, but he would go too. Would go to Placentia or wherever and spend the rest of his days remembering the times back home. What else was there for him to do?

"They's 'll have to drag me," said his Elizabeth with the defiance of a young whippersnapper being sent off to school to repeat a grade.

"They's 'll have to drag me. Without they do, I'll not be going very far. They's 'll have to drag me is all."

She knew in the fall, when the time came, she'd be packing the old clock and her good linen and the quilt her mother gave her for a wedding gift, and she knew in the fall, when the time came, she'd be going too. But she wasn't about to admit it. Not yet. Not until she had to.

In the afternoon we met Stan Ennis. He had heard that Phonse Pittman's son was in and came out to find him. He did find us soon enough and invited us up to his place for a drop o' rum.

"One time," he said, "Phonse was coming over from St. Leonard's to play football, and we was in the same boat together, and I 'ad a bottle o' rum on board, and I passed 'er around to all leaving Phonse out because 'e being the school teacher I didn't know as it's be right to ask 'im to 'ave a drop, and I been mindin' a long time that it weren't right not offerin' 'im a drop, so I wants ye to come up to the 'ouse and 'ave that drop o' rum that yer father should of 'ad that day."

So we went up to Stan's and had hot toddies— boiling water and sugar laced with black rum, "good for what ails ye whether ye be man or beast".

"Don't know what I'll be doing to 'er," Stan said when I asked him what would become of his boat if he had to move off the island.

"Don't allow as I'll be able to sell 'er. Ain't no one'll be left to use 'er anyways far as I can see."

The hot toddy was fit for a king.

"Always was good fish in the bay, leastwise up 'ere. Man could always make a livin' at the fish. But if they takes the plant, my God, what's the use of catchin' 'em."

The next morning, the coastal boat *Petitforte* came in as she was scheduled to, doubling back on the bay run, and we went aboard.

It seemed the whole of Merasheen came to see us off. Men, women and children crowding the small wharf, waving and wishing us well, and saying it was too bad we couldn't stay for the garden party.

Just as the CN boys were pulling in the ropes, preparing to set off, a short stocky man in blue serge, pipe firm in the corner of his mouth, face eroded like a cliff, came over the ramp, walked straight up to me, and, very businesslike, introduced himself.

"I didn't get to have a chat with ye while ye were in which I'm sorry about but I used to know yer father right well when we was young, fished together, first trip for both of us, didn't want it said that Phonse Pittman's boy was in and I never got to say hello to him."

The whistle blew then and he went back over the ramps as suddenly as he had come. Back on the wharf he merged with all the other pipesmoking blue serge that stood hands in pockets waiting for us to shove off.

The people of Merasheen, as warm as we'd always remember them, stood there and waved us away. They waved us past Soldier's Point and out to the Jawbones where we could see crosses almost everywhere upon the cliffs marking the spots where men of the outports had run foul of the sunkers and gone down in the sea that was at once their sustenance and deprivation, their life and their death.

Al Pittman

The Government Game

Come all you young fellows and list while I tell
Of a terrible misfortune that upon me befell.
Centralization they say was the name
But me, I just calls it the Government Game.

My name it don't matter, I'm not young anymore,
But in all of my days I've never been poor.
I've lived a right good life and not felt no shame
Till they made me take part in the Government
    Game.

My home was St. Kyran's, a heavenly place.
It thrived on the fishing of a good hardy race.
But now it will never again be the same
Since they made it a pawn in the Government Game.

Sure, the government paid us for moving away
And leaving our birthplace for a better day's pay.
They said that our poor lives would ne'er be the same
Once we took part in the Government Game.

It's barely five years now since we all moved away
To places more prosperous way down in the Bay
There's not one soul left now, not one who remains,
For they've all become part of the Government Game.

Now St. Kyran's lies there all empty as hell
Except for the graveyards where our dead parents dwell
And the lives of their children are buried in shame,
They lost out while playing the Government Game.

To a place called Placentia, well, some of us went
And in finding a new home our allowances spent,
So for jobs we went looking, we looked all in vain
For the roof had caved in on the Government Game.

It's surely a sad sight, their moving around
Wishing they still lived by the cod fishing ground,
But there's no going back now, there's nothing to gain
Now that they've played in the Government Game.

They tell me our young ones the benefits will see,
But I don't believe it, oh how can that be?
They'll never know nothing but sorrow and shame
For their fathers were part of the Government Game.

Now when my soul leaves me for the heavens above
Take me back to St. Kyran's, the place that I love,
And there on my gravestone right next to my name
Just say I died playing the Government Game.

Al Pittman

I live in Arnold's Cove now, but my home was Harbour Buffet. Owing to the centralization scheme that our famous Joey had on, we had to get out and move to different areas. They didn't like small communities down in Newfoundland and they were going to move us all into bigger areas and there were going to be about four or five jobs for every man that moved. When we moved, we found there were about ten men for every job.

Mac Masters

This Dear And Fine Country —Spina Sanctus

Well, we made it once again, boys!

Winter is over.

Oh, but there is still snow on the ground.

So what? It hasn't got a chance. It is living in jeopardy from day to day. We should pity it because it will soon be ready for the funeral parlour.

It is only a matter of another few paltry weeks and we shall see it disappear into brown and foaming brooks; we shall see the meadows burning green and spangled with little piss-a-beds like tiny yellow suns.

Winter is over.

Oh, but there is still ice in the water.

So what? The globe is turning and nothing can stop it, not even Ottawa. We are revolving into light.

The fisherman tars his boat on the beach and is heated by two suns, one in the sky and another reflected from the water, and the ice on the cliff behind him drips away to a poor skeleton.

It is only a matter of a few more paltry weeks and we shall see the steam rising from the ponds and from the damp ground behind the plough; we shall see the grandmothers sitting out by the doorstep for a few minutes watching the cat; we shall see the small boats a'bustle, piled high with lobster pots in the bow, and the days melting further and further into the night.

Winter is over now.

Praise God and all honor to our forefathers through generations who did never forsake this dear and fine Country.

Ray Guy

The Way God Meant Water To Be

When I was eighteen there wasn't much use hanging around the place no more, the farm where my folks lived, because there just weren't no crops and if I left quietly no one was going to notice and so I got a ride with a cattle buyer and got to Calgary where it was "Move along" and "Keep moving, fellows" and all that, so you could see it wasn't the best place for a single man with no prospects.

One afternoon I walked out west of town and hopped a box car and settled down for the night and when I woke up the train was stopped and I looked out the door and by God, there were the mountains. All around me, pretty as you'd ever like to see, and just over there was a little stream, bouncing along, and I got out and ran over and looked at it. I surely do think it was the first running stream I'd ever seen in August, surely not in southern Saskatchewan, which was my place of abode.

I looked at it, and then I jumped across it and then I jumped back over it and then I stepped into it, and by God, if I didn't sit down in it. Colder than hell but here was water you didn't have to strain and use over again and carry miles to use. This was the way God meant water to be, just running everywhere. Except for the army I never left British Columbia again. That was it. That water.

as told to Barry Broadfoot in *Ten Lost Years*

Acknowledgements

We are grateful to the following for permission to reprint the copyrighted materials. While we have made every effort possible to locate copyright holders, we welcome notice of any errors or omissions and will make certain to correct them in future editions.

"The Creation of Man", adapted by Gary Geddes. By permission of Gary Geddes.

"All the Diamonds in This World", words and music by Bruce Cockburn, copyright © Golden Mountain Music BMI Corp., used by permission. All rights reserved.

"Under the Top of the World", by Joseph MacInnis, by permission of *Maclean's* Magazine.

"Further Arrivals", by Margaret Atwood, from *The Journals of Susanna Moodie* by Margaret Atwood, Oxford University Press.

"Image", from *Tay John* by Howard O'Hagan, reprinted by permission of McClelland and Stewart Limited, Toronto.

"On the Wanapitei", from *Postscript to Adventure* by Charles W. Gordon, reprinted by permission of McClelland and Stewart Limited, Toronto.

"Notes on a Northern Lake" and "Viewing Indian Pictographs at Bon Echo in Eastern Ontario" by John Robert Colombo, reprinted by permission of Peter Martin Associates.

"The Fire Canoe That Was Never Launched" from *The Fire Canoe* by Theodore Barris, reprinted by permission of McClelland and Stewart Limited, Toronto.

"At the Wedding of the Lake" by Rienzi Crusz, reprinted from *The Canadian Forum*, by permission of Rienzi Crusz.

"The Cruise of *The Coot*", from *The Dog Who Wouldn't Be* by Farley Mowat, reprinted by permission of McClelland and Stewart Limited, Toronto.

"In It", from *Happy Enough* by George Johnston, reprinted by permission of Oxford University Press.

"Blue Water Sailing" from *Fair Days Along the Talbert* by Dennis T. Patrick Sears, reprinted by permission of Musson Book Company, 30 Lesmill Road, Don Mills, Ontario.

"The Squall", "Schooner", "Whale Poem", "Lee Side in a Gale", "Offshore Breeze", "Dragging for Traps", "The Sea", and "Boy Fishing at a Pier", copyright © Milton Acorn 1975, reprinted by permission of NC Press Ltd.

"The Wreck of the Edmund Fitzgerald", by Gordon Lightfoot, copyright © Moose Music Ltd., all rights reserved, used by permission of Warner Brothers Music.

"The Edmund Fitzgerald: It's a Year Since She Sank" by Pat Brennan, reprinted by permission of *The Toronto Star*.

"The Marine Excursion of the Knights of Pythias" from *Sunshine Sketches of a Little Town* by Stephen Leacock, reprinted by permission of McClelland and Stewart Limited, Toronto.

"Big Seas" by Terry Cranton, and "Letter from a Small Island" by Randy Lieb, reprinted by permission of *Rural Delivery*.

"Hard Times" from *Beyond the High Hills* by Knud Rasmussen, reprinted by permission of William Collins & World Publishing Company Inc. © copyright 1961.

"Ordeal on an Ice Pan" from *Goodbye Momma* by Tom Moore, reprinted by permission of Breakwater Books Ltd., St. John's, Newfoundland.

"The Annual Seal Test" by permission of Harry Bruce, reprinted from *The Canadian* Magazine.

"Queen of Saanich" by Bert Almon, reprinted by permission of the author from *Whale Sound*, Douglas and McIntyre.

"Killer Whale (Victoria, BC)" by Robin Mathews, reprinted by permission of Steel Rail Publishing.

"If Whales Could Think on Certain Happy Days" by Irving Layton from *The Darkening Fire*, reprinted by permission of McClelland and Stewart Limited, Toronto.

"Saltwater and Tideflats" from *A River Never Sleeps*; "Thoughts Under Water" and "The Death of the Salmon" from *Fishermen's Fall*; "Ripple Rock" from *Saltwater Summer*; all by Roderick Haig-Brown, reprinted by permission of William Collins and Sons & Co. Canada Ltd.

"The Caplin Are In" from *The Foxes of Beachy Cove* by Harold Horwood, copyright © 1967 by Harold Horwood, reprinted by permission of Doubleday & Company, Inc.

"Waiter! . . . There's an Alligator in My Coffee" from *Top-Soil* by Joe Rosenblatt, reprinted by permission of Press Porcepic Ltd.

"Anemones", "Cataract", and "Hermit Crabs" from *A Stone Diary* by Pat Lowther, reprinted by permission of Oxford University Press.

"The Water Molecule", "Ice Worms", "Snowflakes", "Glass of Water", "Water Hole", and "Surface Tension", by permission of Jay Ingram.

"A Minute" from *Toller* by Toller Cranston, reprinted by permission of Gage Publishing Company.

"Jackrabbit" from *Jackrabbit*, reprinted by permission of Collier Macmillan Canada Ltd.

"Peking Captures World Hockey Title" reprinted by permission of *Weekend Magazine*.

"Heart Like a Wheel" by Anna McGarrigle, copyright © by Anna McGarrigle, administered by Garden Court Music, ASCAP. All rights reserved.

from "The Great Bear Lake Meditations" by permission of J. Michael Yates.

"Thaw" by Terry Crawford, reprinted from *Storm Warning*, edited by Al Purdy, McClelland and Stewart Limited, Toronto.

"By the River" from *Spit Delaney's Island* by Jack Hodgins, reprinted by permission of the Macmillan Company of Canada Limited.

"River Road", words by Sylvia Tyson, © Newtonville Music Inc. All rights reserved, reprinted by permission of Chappell & Co. Ltd., Toronto.

"Stefan" by P.K. Page, reprinted by permission of the author from *Vancouver Island Poems*, editor Robert Sward, 1973.

"The Picture" by Sandra Merz, reprinted by permission of the author from *event*, volume 4, number 5.

"The Couch" by W.D. Valgardson, reprinted by permission of the author from *Saturday Evening Post*.

"St. Leonard's Revisited" by Al Pittman, reprinted by permission of the author.

"Death of an Outport" by Al Pittman, reprinted by permission of the author from *Baffles of Wind and Tide*, Breakwater Books Ltd., St. John's, Newfoundland.

"The Government Game" by Al Pittman, reprinted by permission of the author from *For what time I am in this world*, Peter Martin Associates.

"This Dear and Fine Country—Spina Sanctus" by Ray Guy, reprinted by permission of Breakwater Books Ltd., St. John's Newfoundland.

"The Way God Meant Water To Be" from *Ten Lost Years* by Barry Broadfoot, copyright © 1973 by Barry Broadfoot, reprinted by permission of Doubleday & Company Ltd.

Quotations throughout the text courtesy of John Robert Colombo, *Colombo's Concise Canadian Quotations*, Hurtig Publishers; and *Colombo's Little Book of Canadian Proverbs, Graffiti, Limericks and Other Vital Matters*, Hurtig.

Design: Michael Solomon & Tim Wynne-Jones

Illustration Credits

Cover: Ken Steacy

Drawings

Mark Smith 6
Ken Steacy 10, 11, 12, 13
Diana McElroy 17
Selwyn Dewdney, Royal Ontario Museum 20
Leoung O'Young 23, 39, 96
Dale Cummings 34, 35, 44, 47, 48
Rob McIntyre 63, 64, 65, 103, 107
Lissa Calvert 73, 77
Tim Wynne-Jones 80
Libby Masters 83, 84, 85, 86, 87
Kathy Vanderlinden 99

Photographs

Birgitte Neilsen 1, 2, 3
CP Picture Service 42
National Film Board Photothèque 53, 55
Murray Mosher, NFB 62
Christopher Springmann 70
Len Tenisci 84, 85, 86, 87
La Presse 91
Bob Brooks, NFB 118
Crombie McNeill, NFB 120, 125
C. Lund, NFB 122

Notes on the Illustrations

20 Watercolour drawing of an Indian rock painting at Mazinaw Lake, Ontario.
118 Squid jigging, St. John's, Nfld.
120 Cod catch, St. John's, Nfld.
122 *Clockwise from top left:* Happy Adventure, Bonavista Bay, Nfld.; Salvage, Bonavista Bay, Nfld.; Sandy Cove, Nfld.

Canadian Cataloguing in Publication Data

Main entry under title:

Water

(Elements)

ISBN 0-88778-165-9

1. Readers — 1950   I. Carver, Peter, 1936—
II. Series.

PE1121.W38   428'.6   C78-001155-4

© Peter Martin Associates Limited

ALL RIGHTS RESERVED

No part of this book may be reproduced or utilized in any form or by any means, electronic or mechanical, including photocopying, electrostatic copying, recording or by any information storage and retrieval systems without permission in writing from the publisher, except for the quotation of passages by a reviewer in print or through the electronic mass media.

PETER MARTIN ASSOCIATES LIMITED
280 Bloor Street West, Toronto, Ontario M5S 1W1